*Never doubt that a small group of
thoughtful, committed citizens can
change the world: indeed, it's the only thing
that ever has.*

— Margaret Mead

WARNING:
the electricity around you may be hazardous to your health

how to **protect** yourself from **Electromagnetic Fields**

NEW 2nd EDITION

Ellen Sugarman

PRESS
P.O. Box 19-0936
Miami Beach, Florida 33119
1-800-884-6763

Copyright ◊ 1992 by Ellen Sugarman
Copyright ◊ 1998 by Ellen Sugarman

All rights reserved
including the right of reproduction
in whole or in part in any form

Manufactured in the United States of America
1 3 5 7 9 10 8 6 4 2

Library of Congress Cataloging-in-Publication Data is available

ISBN: 0-9661194-0-1

Simon & Schuster edition published 1992

Printed in the USA by

MORRIS PUBLISHING
3212 East Highway 30 • Kearney, NE 68847 • 1-800-650-7888

Acknowledgments

This book benefited from the help and encouragement of many people, including sources in the field too numerous to mention, who gave freely of their time and ideas. I am especially grateful to Connie Parrish, without whom the book could never have been written. I want to thank Sheila Curry for her contribution, and my agent, Al Zuckerman, for his guidance. Thanks are also due to Dr. Robert O. Becker, Dr. David O. Carpenter, Professor Norman N. Goldstein, and Dr. Juan Alcedo for their suggestions and advice. My respect and thanks go out to the many parents of stricken youngsters who graciously shared their private sorrow and public concerns with me during the course of researching the book. And, finally, a special thanks to my parents for all their interest and support.

Table of Contents

Introduction by Dr. David Carpenter ix

Author's Preface xiii

Chapter 1
Electropollution Around Us 15

Chapter 2
What Are EMFs? 32

Chapter 3
Controversy and Cover-up 45

Chapter 4
What the Studies Tell Us 75

Chapter 5
EMFs in Your Community 93

Chapter 6
EMFs at Home 115

Chapter 7
EMFs in the Workplace 134

Chapter 8
RF/MW Radiation 152

Epilogue 169

Appendix A
Major Studies 177

Appendix B
Appliances: What's Safe, What's Not 205

Appendix C
EMF Resources 206

Glossary 208

Index 212

Introduction

The question of whether or not electromagnetic fields cause human disease remains controversial. While some studies show a clear relationship between exposure and incidence of some kind of cancer, others fail to show statistically significant relationships. But recent pronouncements by the National Academy of Sciences and editorials by the New England Journal of Medicine implying that there is no problem, do not necessarily make it so. Both the scientific community and the public remain divided as to whether there is something of concern here.

This book, like the previous one from Ellen Sugarman, provides an account of both the scientific and public debate on this issue. While not everyone from either the scientific or the public community may agree with every interpretation, Ms. Sugarman again presents a provocative account of the evidence and the personalities.

Both the public and scientists have lots of difficulty with issues related to health that are unclear, or in epidemiological jargon, when the risk ratios are low and inconsistent. Everyone likes things to be black or white, where every study gives the same clear result, and where statistical analyses of different studies show either consistently significant or insignificant results. But for many possible health hazards, this is not the case; and this is certainly how it is to date in the study of cancer as a result of exposure to electromagnetic fields. While a number of studies, both of children exposed in homes and of adults exposed at work, have demonstrated statistically significant elevations of leukemia and brain tumors, other studies have not demonstrated any effect at all or have shown only suggestive, not statistically significant, elevations of cancer in the highest exposed persons.

There are many reasons why a real health hazard might not be consistently apparent in different studies. The most obvious cause for inconsistency in the case of electromagnetic fields is exposure assessment, in that it is enormously difficult to determine an individual's total exposure. The studies that have been done to date have considered either only the home exposure from external power lines, or the average exposure from a particular job category. None has been able to examine both residential and occupational exposures, and few have been able to satisfactorily factor in exposure to fields from appliances and other very local sources. All of us go in and out of such exposures many times during a normal day's activities. These exposures unaccounted for may totally obscure a real association, or cause an artificial lowering of the apparent risk ratios. On the other hand, it is also possible to "find" a relation that does not really exist, as a result of what are called confounders. A confounder is some other factor consistently associated with the measured parameter. If the confounder is not recognized, the cause of the disease may be falsely attributed to the measured parameter.

There is another serious problem in how individuals interpret the results of studies of electromagnetic field exposure. Tobacco companies still insist that smoking does not cause cancer. Lead manufacturers in the U.S. convinced Congress that lead added to paint posed no hazard to children at the turn of the century, even though most of the rest of the world recognized lead to be a serious neurotoxin. It took a full fifty years before lead was removed from paint, and still longer before it was removed from gasoline. Chemical companies still deny that pesticides and PCBs cause health effects, in spite of increasing evidence of a host of harmful effects. With regard to electromagnetic fields, the utility industry is powerful and wealthy. While not every scientist who accepts funds from utility interests or holds stock in utilities necessarily allows this to influence his position on the issue, either consciously or unconsciously, there is great danger that financial considerations do influence public positions.

Sir Richard Doll of Oxford University, a prominent epidemiologist who is presently conducting a major study of childhood leukemia and electromagnetic fields in the United Kingdom, has recently urged public health policymakers to carefully evaluate studies that show weak associations, emphasizing that an agent which shows only low risk ratios, if many people are exposed, may have a greater societal importance than an agent with a high risk ratio that affects only a small group. However, Doll remains skeptical on the association between electromagnetic fields and cancer primarily on the basis of the present lack of proof of a biological mechanism explaining the association. The risk ratios reported even in the positive electromagnetic field studies are low, although the difficulties in exposure assessment may cause an underestimate of the true risk ratio. And because all of us are exposed every day to electromagnetic fields, if the low risk ratios are indeed real, the implications for society are very great. The debate over whether there are health effects from electromagnetic fields, and how serious they are, is not over; and this book presents a spirited and personal account of the players, the science, the controversies, and the politics.

My involvement in issues related to EMFs dates from the mid 1980s when I served as the Executive Secretary of the New York State Powerline Project, which among other studies supported the initial research by Dr. David Savitz, then at the University of Colorado, that provided the first confirmation of the conclusion of Wertheimer and Leeper that there was an association between living in homes close to power lines carrying high current and incidence of childhood cancers. Prior to that project I was personally very skeptical that magnetic fields could cause cancer, but this investigation changed my view. While neither I nor other researchers in this area can yet satisfactorily explain the detailed mechanisms that explain the relationship between proximity to power lines or occupational exposure to EMFs and cancer, there is little question but that such a relationship does exist. Certainly it is important to determine exactly what is the mechanism, but in the meantime for each of us as individuals it is most important that we do what we can to reduce our exposures to agents which are harmful. This is why I, along with Ms. Sugarman, advocate "prudent avoidance," the concept of protecting yourself from exposures that are probably but not definitely harmful, if the actions required are neither unreasonably expensive nor disruptive.

> DAVID O. CARPENTER, M.D.
> Dean of the School of Public Health
> University at Albany
> State University of New York
>
> November, 1997

Author's Preface

This book tells the story of an environmental hazard called electromagnetic fields or EMFs. The purpose of the book is to inform and empower. It is only through a genuine understanding of the up-to-date facts about EMFs that you can take steps to protect yourself and your family from dangerous exposure to the fields.

The electromagnetic fields we are concerned with are found everywhere that electricity is in use — around power lines, appliances, office machinery, wall wiring, lights. They've been around since 1882, when Thomas Edison set up the first electric power transmission plant in this country. The tiny little system that lit a few blocks in the middle of New York City was a far cry from what we see today. Currently there are some 350,000 miles of transmission lines crisscrossing the United States.

For a long time, no attention was given to these ubiquitous invisible fields because nobody believed they could be harmful. Then, in the early nineteen seventies, research began to establish some worrisome connections between EMFs and a number of serious health problems, including cancer. But neither the results of the studies nor the warnings of some of the scientists engaged in EMF research were openly communicated to the general public. This book contains evidence of a pervasive and concerned effort — on the part of the U.S. government, the military, electric power companies, and a number of businesses that had a strong vested interest in unrestricted technical development — to cover up the true facts about the potential health risks of EMFs. Instead of warning the public and moving to protect people from everyday exposure to EMFs from the AC (alternating current, an electrical current that changes strength and direction of flow) power system, too many experts and officials have consistently misled the public into believing there was no cause for alarm.

But, little by little, mainly due to bold actions on the part of a few concerned scientists and agency officials, the word about EMFs is finally getting out. Today, a large and growing segment of the American public is aware that the electricity around them may be hazardous to their health — and they want something to be done about it. The EMF question promises to be the main public health battlefield of the nineties, with special interests on one side doing everything they can to keep the truth from the public, and concerned people on the other, clamoring for more information and some sort of regulatory control.

A lot more is known about the hazards of electromagnetic fields than some people would have you believe. The first four chapters of the book tell the EMF story in detail and present a clear, informative guide to electromagnetic fields: what they are, where they are found, what causes them, what the studies tell us about them, and what dangers they pose.

We also know quite a bit about how you can protect yourself from dangerous EMFs without major changes to your lifestyle or going to great expense. Chapters 5, 6, and 7 tell you how to reduce daily EMF exposure in your neighborhood, home, school, and office. Besides advice from the experts, these chapters give examples of EMF problems and how they can be handled. You'll find out how to assess the fields around you, how to discover where they're coming from, what appliances are known to have dangerously high fields, how to redesign your living and work spaces to minimize exposure, and how to maintain your distance from recognized sources of high EMFs.

Chapter 8 is a chapter on radio frequency/microwave (RF/MW) radiation, something that's on everyone's mind these days because of the tremendous growth of the wireless communication industry. With the industry anticipating a cell phone in the hands of every man, woman, and child in this country in a matter of years, the blackout of information on the possible hazards of all this MW radiation is especially worrisome. Before you get in the habit of relying on a cell phone, be sure to read Chapter 8.

The Epilogue talks about what we, as a society, should be doing to protect citizens from EMFs. Energy conservation is of particular importance. Energy can and must be managed differently at the customer end to reduce the electromagnetic fields around all of us.

According to the National Cancer Institute, the rate of childhood cancers in this country has been rising steadily for decades, at the rate of nearly one percent a year. In a matter of two decades, that has resulted in a double-digit increase. The epidemiological studies on childhood cancer warn that we're seeing at least two and a half times the expected number of cases — from 8,000 to 12,000 more cases — because of children's exposures to magnetic fields around them. If this is so, we should all do whatever we can to reduce those exposures — and make our children safe. It is my sincere wish that this book will help us do that.

Chapter 1

Electropollution Around Us

> *If we were to recognize a standard around a high-voltage power line, it would be exceeded everywhere — on the sidewalk, at home, and in the office.*
>
> DR. DAVID CARPENTER, dean of the State University of New York School of Public Health in Albany

In the winter of 1990, teachers at the Slater Elementary School in Fresno, California, began to worry about the high rate of cancer among the staff. First-grade teacher Patricia Berryman recalls, "We originally started to worry when we began to realize that an abnormal amount of cancer cases had been on one side of the school. Then we saw the EPA report on electromagnetic fields and cancer, and we realized that that was the side of the campus near the high-voltage lines. We had some leukemia and brain cancer, just like in the studies. We also had some children with leukemia. Well, we started to research everything we could find." Berryman herself was particularly concerned because "I've spent sixteen years teaching near those power lines."

The teachers began to gather information about cancer cases within the school. "We couldn't trace the whole student body, but we were able to trace the teachers. We went back through all the old yearbooks, made lists of everyone who taught here since the school opened in 1972. We had previous teachers call up and tell us of more cases. Then we made a map and presented it to the parents, to show them where the teachers had been working. Well, their mouths hung open."

WARNING: the electricity around you may be hazardous to your health

Berryman and her group discovered nine cases of cancer in a group of fifty-seven teachers, aides, and lunchroom staff who had worked in two pods (A and B) on the southwest side of the school. That side of the building lies only 110 feet away from a set of two high-voltage power lines — one a 230-kilovolt (kV) line, the other a 115-kV line. One teacher who had worked on that side of the campus for fifteen years had died of brain cancer in 1990, another of melanoma. They also found four cases of children with cancer, but since the state didn't keep any records of the disease in that area until 1987, it's been difficult to trace the children. There were no cases of cancer in teachers who worked in the two pods on the other side of the campus, farther from the lines.

According to the California Department of Health Services, the "conditions" among the school staff that have been diagnosed between 1982 and 1991 include nine invasive cancers (two cases of breast cancer, two cases of uterine cancer, two cases of ovarian cancer, one brain cancer, and two melanomas); one case of skin cancer; one nonmalignant brain tumor; two cases of cervical dysplasia; two cases of precancer of the uterus; three cases of benign breast tumors or cysts; one case of keratosis; one case of sarcoidosis; and a number of basal cell carcinomas. Among the students who attended those pods, the state found three cancer cases and four possible cancer cases, as well as two nonmalignant tumors and one cervical dysplasia.

Besides worrying about cancer, some of the teachers were also concerned that the electromagnetic fields had bad effects on the children's learning and behavior. "For years, we had noticed the behavior of the children in those classrooms. They couldn't pay attention, they fidgeted, they just couldn't keep their feet on the floor. We kept a journal while we were still there and knew we were going to be moving to the trailers. In those days, we really thought first graders simply could not sit still. It all changed when we moved away from those lines."

When word of the problem started to get around in the spring of 1991, many teachers refused to work in the classrooms near the power lines. Fourteen teachers requested transfers to other schools. There was even talk of a teachers' strike. But, still, nobody seemed to listen to the teachers until the parents became involved. Patricia Berryman explains, "I remember a parent came up to me one day and asked, 'What can we do to help?' The parents picketed and invited all the news media to attend. After that we got immediate action."

More than a hundred parents and children picketed Slater in May, 1991. They carried placards that read "SAVE THE CHILDREN" and "PRECAUTIONARY MEASURES MAKE COMMON SENSE." The demonstration kicked off a schoolwide boycott that had been organized by parents like Lynn Stetson, who is currently president of the Parent Teachers Association and one of the co-chairpeople of an ongoing EMF task force at Slater. The parents demanded that classrooms and a portion of the playground near the power lines be closed.

The Fresno Unified School District responded to their demands immediately by closing the ten classrooms, placing the children in ten portable trailers on the other side of the campus at a monthly cost of $695 per trailer, and closing off an area of the playground that was nearest the power lines.

A May 17 memo to the superintendent from Fresno Unified School District Chief Financial Officer Cathi Vogel recommended the changes, stating, "Although evidence is inconclusive regarding the potential hazards of power lines, the concerns of Slater staff, students, and parents are impacting the instructional program and must be addressed immediately."

Donald Beauregard, Fresno area administrator for the FUSD, said: "We told everyone it was a temporary thing. We had to do something. We had teachers who were refusing to teach in those classrooms. This was about human health, teachers, and kids. There was a tremendous amount of fear and, as you realize, fear can interfere a lot with the overall educational process. Something that bothers me is, right now, we still have schools around the county that are being built right near transmission lines."

Beauregard serves on the EMF subcommittee with Berryman, teacher Sandra Craft, and Lynn Stetson. The committee meets regularly to grapple with the overall issue of whether or not there is a danger from the power lines and, if so, what to do about it. Stetson refers to the committee as "our little group," adding that she always feels they are the "underdogs" at public meetings.

Berryman concurs. "It's terrible. There are only three of us and Mr. Beauregard. On the other side of the table PG&E (Pacific Gas and Electric) has all these engineers who keep telling us there's nothing to worry about from the power lines. Then the state health department keeps telling us there's no funds to do a study of our school. The state brought in epidemiologists from the county health department. The first was a Dr. Stallworth and he just got up and told us we were all wrong, there was nothing to it. Well, after that one meeting, we never saw him again. They told us his report got lost in the computer. Now we have a woman named Betty Carmona. She tells us she doesn't know much more about this than we do ourselves, but she seems sympathetic."

Dr. Eva Glazier, the state health department epidemiologist in charge of the Slater cancer study, said her goal was "to determine if the number of cancer cases is significant." The problem is, she explained, that the teachers "have no way of actually knowing where anyone taught, so we have to check that out." When asked how they were doing that, she admitted that

WARNING: the electricity around you may be hazardous to your health

the Fresno School District hadn't kept any records of teachers' room assignments. So the health department apparently will be using the same sort of "informal data collection" methods that the Slater parents and teachers used — that is, asking people what rooms they worked in.

When questioned as to her opinion about the controversy, Dr. Glazier said, "I know they're (the classrooms) very close to the lines. Once the possible dangers have been pointed out and you stand there and look at those lines — it's creepy."

The concerned parents and teachers see the trailers as a temporary first step and have continued to lobby the district to build permanent portable classrooms to replace the trailers, to move some of the grades to another school, and to replace playground equipment. They also want counselors available at the school to deal with the children's cancer fears.

According to Lynn Stetson, there has been a great deal of conflict in the Slater community about whether or not the power lines are dangerous — and what to do if they are. She feels that part of the problem may be concern about property values. "At the beginning, the school district and the community were not behind the teachers," she explained. "When the parents boycotted the school, that's when people began to listen. We feel we really do have a genuine problem because, even though PG&E keeps trying to downplay the problem, they have responded seriously by loaning us meters to measure the fields and meeting with us."

There was strong opposition to the protests and the issue created factions in the community. One of the original organizers was certain her phone was tapped and a break-in occurred at her house when she was directly involved in the EMF committee. Early in 1992, PG&E went to the League of Women Voters and suggested they host a panel about the issue. Stetson says, "We'll be there on the panel, but we haven't been able to find out who PG&E is bringing. They keep coming up with experts who tell the public, 'There's nothing to this. We don't have to do anything.' We feel as though PG&E was setting us up."

Around that time, the parents were instrumental in getting the school district to bring in a second, independent tester. (Fields had been measured previously by PG&E and a group called Magnetic Measurements.) The district hired Karl Riley of ELF Magnetic Surveys in Berkeley, a company that specializes in locating the sources of magnetic fields and providing mitigation action.

Riley recalls that he was a bit taken aback at the welcoming committee when he got to the school: Beauregard, two assistant superintendents, the school principal, Stetson, Berryman, two other parents, Pam Long of Magnetic Measurements, a PG&E spokesperson, and two school electricians.

Riley found high AC (alternating current) magnetic fields — 24 mG (milligauss) in the center of one room, 108 mG on the floor of another room — in the classrooms that had been vacated because of high magnetic field readings in the past. In each of the classrooms, he was also able to locate the source of the high fields — improperly installed wall wiring. With the help of the school electricians, he was able to correct the problem and reduce the magnetic fields to less that 1 mG. (A gauss is a measuring unit for magnetic field strength: 1,000 milligauss [mG] equal 1 gauss.) Riley explains:

> I determined that the high magnetic fields were due to the particular way the switch currents were wired, which allowed conduits to carry either a hot wire without its paired neutral or a current-carrying neutral without its paired hot wire. In electrical terminology, the ungrounded conductor and the grounded conductor had been separated into separate conductors. When pairs of current-carrying wires are separated, they no longer mutually cancel each other's magnetic fields, so their fields are free to spread out. Hence the high field measurements. The problem was new to the electrical maintenance men, but the man working with me did in good time understand what we were attempting to achieve (hot and neutral together) and he was quick to make the simple changes in the switch box to achieve the pairings which result in balanced magnetic fields with no net current.

Eventually, the EMF situation at Slater School was fixed to everyone's satisfaction, according to Don Beuregard. "After we went through all the dialogue, the district chose to keep those particular pods for storage, not pupil use. They put in portable buildings farther from the power lines for the classes they displaced. That seems to make everyone happy."

The Amadors have just moved into a rambling, five-year-old northern California ranch house. They haven't quite finished unpacking — the garage is half full of boxes — although all the essentials for family living are comfortably in place. Laura and Art Amador have three boys: five-year-old twins and eight-year-old Arthur.

The neat little bathroom off the master bedroom has flowered wallpaper; a set of pink, lace-trimmed towels; and a ceramic lamp with an eyelet shade. There's a basket of assorted guest soaps still in its pink cellophane wrapper on a shelf, and a magazine rack full of ladies' magazines. But,

WARNING: the electricity around you may be hazardous to your health

amidst the general coziness of the decor, there's an opened paperback book lying face down on the corner of the sink — *Killing Cancer: The Jason Williams Story* (Smythe, Vinton, 1980). It's a chilling reminder that beneath this noisy, cheery atmosphere winds a dark current. Arthur Amador has cancer. Arthur got sick in late 1989, a few months after the Amadors moved into a costly new house in San Ramon. Laura recalls, "Over a period of six weeks, Arthur developed eight tumors on his neck that eventually stuck out farther than his ears. You could see them growing. We went around to doctors, but at first they kept telling us he had mono; they were just swollen glands. But I said, 'I think my son has cancer.' Finally they were able to diagnose him at Kaiser in Oakland. It was non-Hodgkin's lymphoma."

Cancer of the lymph nodes. "The tumors had already grown to where he was stage four — the cancer was everywhere. . . . When people say, well it's only a small number of children affected, I think about how much Arthur suffered and what a terrible disease it is." Arthur endured an eighteen-month course of therapy, including chemotherapy, which cost $700,000.

Medical science prevailed and Arthur has been in remission for more than two years. Today, the blond boy looks perfectly all right. But you can see the fear in his mother's eyes as she says, "He's in remission; he licked it."

Laura Amador is a petite, perky young blond woman whose energetic appearance belies her recent struggle. Laura works part time as a supermarket checker. Her husband, Art, builds custom kitchens. Laura believes Arthur got sick because of the electromagnetic fields emitted by power lines near the San Ramon house and Arthur's school:

> Once you find out your child has cancer — after you get over the shock — you want to find out why. I started reading. I saw a Family Circle article by Paul Brodeur. It told about kids getting cancer from electromagnetic radiation from power lines. That was the first time I'd heard of EMFs.
>
> Then I realized our house was right near the high-voltage lines that run over the Iron Horse Trail. We walked right under them every time I took Arthur to school. He used to ride under them on his bike.

Mounted on 80-foot steel towers, an array of 230-kV transmission lines and lower-voltage primary distribution lines run for miles above the Iron Horse hiking trail. The lines also run near Monte Video School, which Arthur used to attend, two other elementary schools, two day-care centers, and a Little League playing field.

A number of well-accepted studies have reported that exposure to power line magnetic fields greater than 2.5 mG more than doubles the risk of childhood cancer. In fact, some experts believe 15 to 20 percent of the childhood cancer in the United States today is due to EMF exposure.

When the magnetic fields from the Iron Horse lines were measured, 105-mG fields were found on the trail under the lines, 1.5- to 6-mG fields were measured near homes 50 feet away, 80- to 102-mG fields were found at the crosswalk to Monte Video school, and 20- to 60-mG fields were measured in the bleachers at the Little League field.

John Hannes, a neighbor who lived 15 feet away from the lines and has since moved, measured 40mG fields in his home last year. At the time, the power company assured Hannes this was nothing to worry about. So did the city. But the Hanneses were worried. They were also concerned about plans to build a Little League field under the transmission lines. They wrote to the City of San Ramon, expressing their concerns.

On May 23, 1991, Hannes received a letter from Jeff Eorio, San Ramon's parks director, that said, in part: "The EMF task force and the Parks and Community Services Commission have reviewed your request to relocate the new ball field . . . farther away from the Pacific Gas and Electric distribution line . . . and determined that there are insufficient data available concerning the potential ill health effects of EMFs to make a responsible decision."

The City of San Ramon built a children's baseball field only 50 feet away from the high-current distribution lines, in fields that measured from 20 to 60 mG, which is thirty times the level that, according to the latest studies, has been linked to increases in childhood cancer.

Laura Amador decided to share her concerns about the dangers of EMFs with her San Ramon community when she learned that one of the neighbors was trying to get a cement walkway to the school installed right under those lines. "I thought I'd better to go the city council and warn them about the danger. I had no idea what a can of worms I'd be opening."

Laura went before the San Ramon City Council and described her concerns. In response, the mayor organized a meeting between the Amadors, a representative from Pacific Gas and Electric, and two city staffers to discuss the problem. At the meeting, Laura says PG&E told them there was no problem:

> They said there was nothing to worry about and they weren't going to do anything to change the lines until the government set standards that required them to do so. Then the city planner said the council was going to take a cautious, conservative

stance. Later, in private, he told us they intended to do nothing about it and we should do likewise. We should stop worrying people.

But Laura Amador didn't keep quiet. Instead, she formed an organization called Citizens Concerned About EMFs. Members did research on the connection between EMFs and cancer and distributed packets of information to the community. They called meetings and invited experts to speak on the subject. One such expert was Raymond P. Neutra, chief of the epidemiological studies surveillance section of the California State Department of Health, who spoke before a group of thirty concerned homeowners. Laura says she thought Neutra "leaned toward telling us they're finding unexpected things that are dangerous to health, but nobody seemed to know how it works."

What Dr. Neutra didn't tell his San Ramon audience was that, around the same time, his department had been instrumental in helping another group of concerned parents in Montecito, California, who had similar concerns about power lines adjacent to their children's school.

Montecito is a wealthy community near Santa Barbara, in southern California. The parents believed that a high incidence of leukemias and lymphomas among children who had attended the Montecito Union School was due to electromagnetic fields from an electrical substation and a 66,000-V (volt) transmission line near the school. The Department of Health had recorded magnetic field levels as high as 12 mG on the school playground, but it initially tried to play down the health threat. However, the parents hired an independent engineering firm called Enertech to take a new set of magnetic field readings, and Enertech came up with levels of 600 to 1,000 mG near the transformer and 18 mG in one of the classrooms near the power line. (One reason for the wide disparity in the readings had to do with the fact that magnetic field levels change depending on the amount of current in use: Peak field levels occur at times of peak power usage. The State Department of Health took field measurements on a Sunday afternoon, and Enertech measured the fields on a weekday, when people would be expected to be home using more electricity.)

In the spring of 1990, state epidemiologists reached a compromise solution, accepting the school board task force's recommendation to close classrooms nearest the lines and rope off the areas of the school playground with high magnetic fields. (The utility would not entertain the notion of relocating the offending lines, although many parents wanted them to do so.) According to Richard Kreutzer, an epidemiologist with the toxicology section, the department points to Montecito as a model of successful managing of electromagnetic field concerns:

We consider it pretty much a success story. How can you manage an ambiguous risk? In general, numbers were found to be above 2.5 mG. Areas on the playground had to be walled off, painted with warnings. Behavior had to be shifted so far as use of some classrooms. Some relatively inexpensive and simple actions were undertaken. I think Montecito is a case study of a reasonable way to deal with this concern in the short term.

In San Ramon, however, things didn't go as well. The Citizens Concerned About EMFs continued their lobbying, held meetings, and got a lot of publicity, but they couldn't get the rest of the community behind them. (As in Montecito — where the concerned parents had prevailed — property owners in San Ramon were worried about a threat to real estate values and tried to halt the investigation.) Laura intensified her crusade to warn others about the hazards of the Iron Horse line, speaking to reporters and appearing on television. She began to get hate calls warning her to stop causing trouble. She called for a meeting of the PTA at Arthur's school. But the day before it was scheduled, an attorney on the city council phoned a number of key PTA members to inform them that there was nothing wrong and that Laura Amador "hadn't been thinking straight since her boy got sick." The PTA canceled the meeting.

"That's when I got out of the group," Laura explains.

"I got so discouraged. I know I got a lot of people educated about the problem. But I would have liked to see those lines moved." Instead, the Amadors moved. Besides the pressure they were feeling from their opponents in the community, they had another reason for relocating. Arthur was afraid of the power lines. "It got so he couldn't sleep. He could see the lines from his bedroom window. He used to come into our room night after night," Laura recalls. "We picked this house very carefully. There are no lines anywhere. The fields in every room are low."

Laura Amador is keeping a low profile on this issue these days, but Citizens Concerned About EMFs haven't given up the fight. The group is still active in northern California as a clearinghouse for EMF information.

When citizens in Middletown township in New Jersey joined forces to stop a high-voltage power line that Jersey Central Power and Light (JCPL) wanted to build, the scenario was quite different — the local government was fully behind them. In fact, the township itself became involved in legal action against the power company. (More and more, local governments around the country are getting involved in EMF issues, going to court in place of the individual citizens who originally raised these concerns.)

WARNING: the electricity around you may be hazardous to your health

Jersey Central Power and Light wanted to erect a 10-mile-long 230-kV transmission line along an existing railway right-of-way (ROW) within 15 feet of a residential area. Mayor Rosemary Peters recalls:

> Visually, the towers would have been a real blight. Part of the line would have gone right through our historic district, one of the prettiest sections of the township. And of course there was the health issue as well. I had read enough to feel there definitely was a problem. . . . What made it work was a governing body that joined in the fight, a citizens' group that was credible and extremely well organized, and a crackerjack attorney who lived here and was used to dealing with the federal court system. But it's only going to happen where a town or a group has the money to fight it legally. It cost us $300,000 and we virtually did it on a shoestring. The utility brought in a stable of witnesses, essentially the same dog-and-pony act they trot out every time there's a case like this anywhere in the country.

It all began in March, 1989, when Jersey Central outlined the proposal at a public meeting and met with resistance from an organization called RAGE (Residents Against Giant Electric). There had already been an EMF controversy in a nearby town called Little Silver, so people in the area had been warned about the dangers of EMFs. Besides that, Middletown township has a large number of scientists and people with technical knowledge. Mayor Peters says that helped.

One of the founders of RAGE was Barbara Iannucci. Iannucci's son had cancer, and she was worried about the potential dangers of exposure to high electromagnetic fields. She decided to call up some of the experts. "I talked to Nancy Wertheimer and David Savitz (authors of two key epidemiological studies that linked magnetic field exposure to childhood cancer). I said, 'I have a child with cancer. Will he be affected badly?' They said, 'Right now it's an inconclusive health issue, but your thoughts are not ill grounded.'" That was enough for Barbara to decide to carry the fight to the utility.

But Jersey Central bypassed the usual hearing procedure and brought a petition before the state regulatory commission, saying that the line was going to be opposed. Under New Jersey law, this made it necessary for the issue to be dealt with in the courts. Iannucci credits attorney Mel Greenberg, of the Newark, New Jersey, firm of Greenberg and Epstein, with handling "all the details and procedures" for the citizens' group because he was experienced in the federal courts. Greenberg was also able to demolish the utility's case for need, a critical question in the siting of new power facilities. Usually, regulatory bodies take a utility's word as to the need of a new line. Greenberg explains:

> I have been told that among the power companies this case has caused some consternation. I think it provides a case study in how power companies put on their blinders and assume that with their money and experts they'll win. The companies come in like 800-pound gorillas and just assume they're going to win. They're not treating this as an environmental issue, which it is. They're not showing any sensitivity about how people feel about EMFs. And so far as their experts are concerned, if you read the transcripts, some of their testimony is a joke. These folks have been testifying one way for years, but after the OTA (Office of Technology Assessment) report came out, they had two choices. They could change what they were saying, and prove themselves wrong, or keep saying the same thing, that there was no evidence of health risks from EMFs. Well, they chose to maintain the same stance. (The 1989 OTA report Greenberg is referring to is the first government document stating that the possibility of EMF health risks could no longer be dismissed.)

RAGE was also successful in organizing the community to oppose the new transmission line. The group persuaded five hundred people to come out for a key public hearing on the matter, which may be some kind of a record.

"We felt the drumbeat of public opinion," recalls Peters, describing the two-night, ten-hour-long meeting. RAGE and Greenberg devised what proved to be an effective strategy and coordinated the public testimony at the meeting. Instead of everyone getting up and saying the same thing, that they didn't want the line, each person spoke on a different facet of the problem. "It showed the judge that we had an educated public. It was democracy at its most beautiful. We were being constructive, not obstructive," says Greenberg. "The judge constantly referred to the public meeting."

Next, Iannucci and Greenberg spoke to a member of the governor's staff about the problem, which prompted the New Jersey Public Advocate to come into the case on behalf of the Township. That sent a signal to the utility. When Jersey Power and Light was overruled on Premise One — need — the utility withdrew its proposal.

Dr. Andrew Marino, an EMF researcher from Louisiana State University Medical School who was one of the township's expert witnesses, still has a problem with the outcome of the case:

WARNING: the electricity around you may be hazardous to your health

There are many instances now in which town boards raise the issue, like they did in Middletown, and the other sides back off to avoid a precedence situation. Which is one of the nasty sides of the American legal system. If a defendant backs off, that permits information of great societal import to be hidden from the public. Then the next such plaintiff has to reinvent the wheel.

A group of New York State property owners has spent nearly eight years reinventing that wheel. The Filipowski property sits on a quite rural road in Middletown, New York. The Filipowkis have lived in their centuries-old farmhouse for nearly fifty years. John Filipowski is a retired Middletown building inspector, sewage treatment plant operator, and a private developer. The Filipowskis raised a family in this house, John's father spent his last days here with them, and right now their twenty-one-year-old daughter and her husband live in the little apartment in the front that John built on for his dad. The wood-paneled living room still has its original wide-plank oak floors and a central stone fireplace.

Despite the high temperature, Mrs. Filipowski is positioned over the kitchen stove, baking a dozen of her special zucchini pies. She's a small, quiet counterpart to her robust, gregarious husband. The aroma of baking fills the living room, as the Filipowskis sit surrounded by their "collections" — his beer steins, her Hummel figures — and explain how drastically their lives have changed since the Marcy-South power line came along in 1985.

"I have twenty-three windows in this house," John Filipowski says, sweeping his hand to describe a 180-degree view. "And I can see that line from fourteen of those windows."

He's talking about two 345-kV transmission lines mounted on steel towers that make a dogleg turn at one corner of his property only 400 feet from the house. When you stand on the front lawn, you can hear the snap, crackle, and pop of the electric field. Filipowski says he used to come out and pick night crawlers off the lawn, but since the line went in, the worms are gone. So are the flying red squirrels that used to inhabit the old barn.

Filipowski is worried about the electromagnetic field from that power line. He has sometimes been photographed at night standing under the lines holding two 40-watt fluorescent bulbs — unplugged — that glow eerily in the dark as they suck up power from the field. The 207-mile-long New York Power Authority line that brings hydroelectric power from Quebec and Ontario to southern New York state was turned on, in spite of the objections of many locals, in 1985, and is still the subject of an ongoing dispute.

Filipowski refers to the line as "a death sentence." He says, "We're deathly afraid. This house is only 400 feet from the line and we don't know what the effects may be. Our daughter's recently married and they live with us. We're afraid for them to have children."

Besides his concerns about his family's health, Filipowski says New York Power Authority has stolen their future as well, by halting his development project — a ninety-eight-lot subdivision, a lake, and a clubhouse. He had completed half the houses and had just begun dredging the lake when the line was built. "I worked eighteen hours a day, two jobs my whole life, to be able to do this," he complains. "They've ruined it and they're unwilling to compensate me. No one's going to buy a new home next to this power line. I doubt if I could sell my own house if I wanted to."

John Filipowski isn't alone in that feeling. In 1988, he joined fifty-nine other landowners in Orange and Sullivan counties in a class action lawsuit against NYPA for devaluing their property by creating a "cancer corridor" where people are afraid to live.

> Land, enhancements, improvements indirectly appropriated as a de facto permanent taking by creating a corridor 2,400 feet wide . . . within which corridor EXTENSIVE AND DANGEROUS LEVELS OF ELECTROMAGNETIC FIELDS and other pollutants will emanate from the transmission line causing biological effects resulting in health hazards and a carcinogenic atmosphere, rendering said area unsafe to reside thereon, cancer phobia and being otherwise dangerous to the health of humans, flora, and fauna.

The lawsuit is being handled by Gurda, Gurda, Lynch and Smith of Middletown, New York. (Another thirty-nine property owners in three other counties are being represented in a similar action by Margarettville attorney Joseph C. Sharpiro.) Attorney Michael Gurda III says the lawsuit charges that the construction of the transmission lines constitutes a taking of the land without fair remuneration, as well as diminishing its value. (A unique legal situation exists in New York State, where the NYPA, as an arm of the state, has sovereign immunity: Personal injury lawsuits aren't permitted as they are in many other states.)

"The power company says this is an easement, not a taking," Gurda explains. "What that means in New York State is, the power company has the absolute right to do what they're doing — but the landowner still has the absolute right to pay taxes on the land."

WARNING: the electricity around you may be hazardous to your health

Originally, Gurda and his father, Michael Gurda II handled the Marcy-South case together; recently, he more or less inherited it when his father became ill. Gurda says, "My father always said, 'If there's a wrong, there must be a remedy.' He took this case because he felt there was something wrong here."

The case has been a drain on the relatively small firm. "The Power Authority has spent $2.4 million to defend this — $667,000 on expert witnesses alone. The plaintiffs have two attorneys. The power company has had as many as twenty-six attorneys in the courtroom at one time. To tell you the truth, I really don't have the unlimited resources to continue to fight them. It costs a fortune. I have sixty clients."

This lawsuit was the first in the United States to take on the twin issues of the health hazards of the electromagnetic fields and their effect on property values. Thus, it should have important precedent-setting ramifications. Today similar litigations are being brought all over the country. But Gurda explains there are inconsistencies in these cases "because there are different rules in different states. For one thing, there are different rules for the personal injury litigation according to what state it's tried in. This results in a different justice for different people. This is actually a national issue that's being treated differently by every state. States can't handle it, communities can't handle it — it's got to become a national issue. This is why I believe the federal courts will have to become involved." Gurda also predicts that as cases begin to settle around the country, we're going to see a lot of gag orders to conceal the terms of the settlements.

The upstate New York plaintiffs also have another gripe: The affected communities got more compensation from the utility than individual landowners did. The NYPA wears two hats in New York State: It's both a state agency and a private entity. As an arm of the state, it has the shield of the state when it needs it (for example, lawsuits against it are heard in state appeals courts, in front of judges who are appointed by the governor and paid by the state, rather than going before juries). But it can still act with the same rights as private companies, such as when New York Power Authority distributed grants to municipalities in the path of the Marcy-South line. Communities received grants based on the amount of line passing through them, at the rate of $5,000 per mile. The only stipulation was that the money be used to benefit the public. Opponents of the line, who call these grants "sin money," say this as a form of bribery, an obvious way of greasing the wheels of local governments besieged with citizens' protests against the line.

In September, 1989, Court of Claims Judge Peter C. McCabe ordered New York Power Authority to pay plaintiff Donald Zappavigna $41,215 on the basis that the sight and noise of the line had indeed devalued his dairy farm. The judge threw out the cancer-phobia part of the claim because he didn't feel there was enough proof that electromagnetic fields cause cancer. Even so, the decision represented a first and, with another 142 plaintiffs in line, NYPA immediately filed an appeal. Ironically, due to the increasing information in the last two years linking EMFs to cancer, that appeal could now make it possible for Gurda to resurrect the cancer issue in his cross-appeal.

Meanwhile, Gurda says he understands the judge's decision:

> The court is a practical entity. It has to be. The judge was trying to be reasonable. . . . What I find fault with is this public service commission. They should have a different standard than the judicial branch. Their job should be to protect the public without the restraints of the judicial branch. If in fact there is a general effect to health, then the power companies need to start handling it properly. [The New York State Public Service Commission approved the power line in 1985, over strong objections from the public and a number of expert witnesses who testified against the project.]

As Gurda points out, litigation also serves the purpose of making people aware: "It's like this. Someone gets $5 million for cancer caused by power lines. Then everyone's going to say, 'Hey, EMFs cause cancer.'"

Medina, Washington, is a residential community of 6,000 souls — one of them Bill Gates — outside Seattle. Like other communities around the country, in the last few years Medina has experienced what City Attorney Kirk R. Wines calls a "significant increase" in the demand for wireless communication facilities within the city. In response to citizens' concerns about potential health hazards, aesthetic effects, and property devaluation from these facilities, the city imposed the first of three moratoriums on permits for communications facilities in February of 1996. The city fathers wanted time to study "a large amount of new information that was not provided by the wireless industry." Siting the facilities without adequate study, the city contended, was likely to result in adverse effects on the health and safety of its citizens. Furthermore, the city wanted to wait for the rules the FCC was developing regarding the environmental effects of radiofrequency (RF) emissions which might impact the placement or configuration of wireless facilities.

The Medina approach is a model for communities with similar concerns; the moratorium defined the problem and set up a number of clear goals, among them:

- To study... the actual effects, if any, of wireless antennas upon the values of adjacent residential properties.
- To consider... methods for testing actual levels of microwave (MW) or radiofrequency emissions.
- To study what other cities, towns, and counties have done to address concerns relating to wireless communications facilities.
- To study whether the Federal Telecommunications Act and the Federal Aviation Administration regulations... have effectively preempted local regulation of microwave and radiofrequency emissions and whether such preemption is constitutional... (and) to seek a formal opinion from the Office of the Attorney General as to the effect of the FAA regulations on the ability of a municipality to regulate MW and RF emissions.
- To consider revisions to the zoning code to address the concerns that have been raised by the community.

While Medina held the wireless industry at bay, the city conducted meetings and workshops to gather information. Hundreds of people signed petitions supporting the moratoriums and asking for revisions in the zoning code.

"We took time to build an adequate record," explained Wines. "That was the key: to develop substantial evidence on the local record." Along with expert testimony, various citizens were assigned a topic to research thoroughly and testify on.

In April of 1996, Spring Spectrum applied for a permit for a 100-foot antenna. Medina denied it. Sprint came back with an application for a 75-foot tower. Medina turned that down too. Sprint filed a motion for a preliminary injunction declaring the moratorium invalid under the Telecommunications Act, which provides that a local government regulation "shall not prohibit the provision of wireless services."

Judge William Dyer denied the injunction, ruling that the moratorium was a "bona fide effort to act carefully in a field with rapidly evolving technology." The judge also found that Sprint's argument incorrectly implied "that federal law bars ANY rejection by local government of a wireless communications provider's application for a zoning variance. That is contrary to the Federal Telecommunications Act, which EXPRESSLY PRESERVES CITY AND STATE ZONING AUTHORITY.

Sprint appealed the decision, then withdrew the appeal.

Meanwhile, Medina drafted another model document — an ordinance covering Wireless Communication Facilities. Noting that "traditional zoning has concentrated on the preservation of property values," Medina addressed citizen concerns by setting a 35-foot height limit for towers, requiring the companies to use so-called "concealment or stealth technology" to hide the antennas, and requiring 500-foot setbacks from the nearest residence. Stating that "income is not even a consideration for approving or denying an application," Medina set a $5,000 permit fee, on the basis that fees greater than $5,000 were being charged around the nation. Annual licenses could be renewed — but after year five, the applicant must reapply "as if for a new facility."

According to the ordinance, "despite the fact that the city has approved one or more facilities in the past doesn't mean it must approve a similar facility for each potential provider." Finally, applicants must demonstrate need and prove the facility will provide services primarily for residents of Medina.

Chapter 2

What are EMFs?

We live in an electrical environment. As we go about our business, just about everything we do gets an assist from electricity. At every point in our daily routine, we see evidence of this dependence on electrical power. We wake in the morning to the sound of our digital clock radio. Moments later, we use an electric shaver, an electric hair dryer, even an electric toothbrush. Next, we move into the kitchen where we stick some bread into the toaster and turn on the coffee maker. Those of us who depend on public transportation may ride on electric subways or buses. From the moment we arrive at the office, most of us spend our workday in tandem with any number of clever electric devices that make the work go quicker and easier: computers, copying machines, fax machines, even the overhead fluorescent lighting and the soda machine. When we return home, we may use an electric oven or microwave and, finally, settle down for a few hours of relaxation in front of the TV or plugged into the stereo headphones. From the second we wake up to the time we go to sleep, nearly everything we see or touch or use has some sort of electrical component.

THE ELECTRIC POWER SYSTEM

The *electric power system* is a giant power grid crisscrossing the country and providing us with the energy we need to light our homes, offices, and factories and run all those machines and home appliances. Humans have been able to harness electricity for more than a century, ever since Thomas Edison started the first electric power company in New York in 1882 and Nikola Tesla developed the first electric power system in 1893, amazing the world by turning night into day at the Chicago World's Fair. From that day forward, the world has become a very different place. Americans have been enamored with electric power — never realizing that hidden in its magic is a dark seed of destruction.

What are EMFs?

In the words of David E. Nye, in his book, *Electrifying America* (MIT Press, 1990):

> There was never a time when ordinary Americans understood electricity in purely functional terms; they have always responded to it with a touch of wonder and ascribed to it symbolic dimensions. . . . The electrification of America is thus far more than the story of inventions and corporations; it involves a popular absorption in the potentialities for personal and social transformation. As America electrified, in the imagination it became electrifying.

THE BASICS OF ELECTRICITY

The *electric power system* is made up of a series of components that produce electrical power, then take it from the source where it's produced and move it along power lines to the homes and offices of the power company's end users, or customers, like yourself.

At different points along the way, the voltage is either increased (stepped up) or reduced (stepped down) by *transformers*. Power lines themselves differ according to how much power they are designed to carry, with thicker lines being able to handle more juice. While the line voltages in a system remain constant, currents vary according to customers' use. Peak currents occur at times when the largest amount of electricity is in use.

Generators, high-voltage transformers and *substations*, and *high-voltage transmission lines* all deal with the high-voltage end of the power delivery system. Taken together, they're referred to as *transmission components* and their job is to handle the power as it's transported from the generating plant to the region where it's going to be used.

Once the user community is reached, the distribution piece of the system comes into play. *The distribution lines* deliver electricity to individual homes and businesses. In general, but not always, distribution lines handle lower voltages and lower currents than transmission lines do, with one important exception. DURING PEAK POWER USE, THE AMOUNT OF CURRENT TRAVELING OVER DISTRIBUTION LINES CAN ACTUALLY BE AS HIGH AS THE CURRENT IN A TRANSMISSION LINE. Figure 2-A diagrams how electricity is moved from place to place.

Power of about 20,000 V (20 kV) is generated in gigantic power stations usually located in sparsely populated rural areas. Huge hydroelectric power stations harness and convert the power of our mighty rivers at places like Niagara Falls in New York or Hoover Dam on the Colorado River, in the

WARNING: the electricity around you may be hazardous to your health

southwest. Electrical power is also generated at coal, oil, and nuclear power plants all over the country.

The voltage produced at these plants is immediately increased, by large step-up transformers, to voltages in the range of 69 to 750 kV. Then it's ready to be moved long distances over high-voltage power lines (able to carry 500 to 750 kV of electricity) to user communities. These high-voltage lines have to be mounted on specially designed 50-meter-high metal or wooden structures.

You'll probably recognize the high-voltage towers in Figure 2-A. They're the towers you see across the countryside. They aren't easily confused with anything else in the electric power landscape. You've probably also seen a transformer yard or distribution substations. They have an unmistakable intergalactic look to them. Step-down substations, which are found in many neighborhoods, are a little more difficult to spot because they can be disguised to resemble ordinary neighborhood buildings, such as houses.

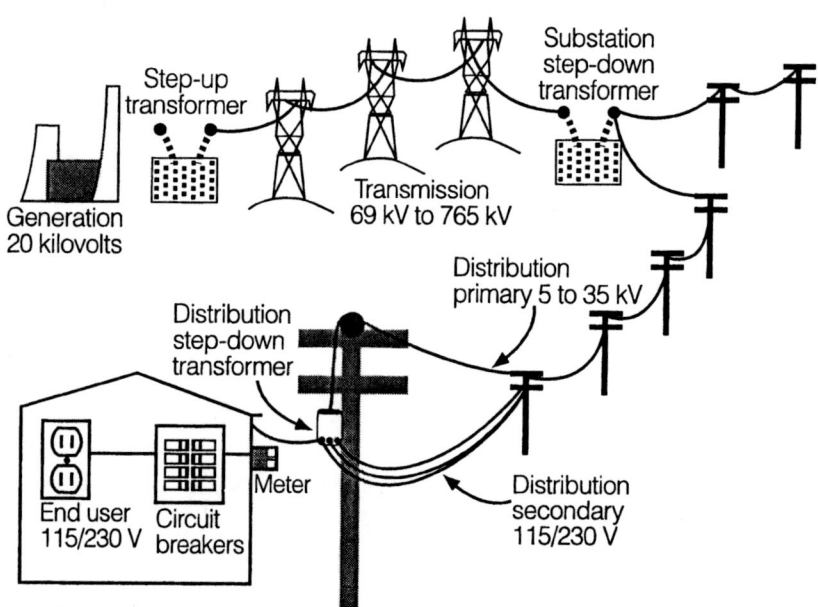

Figure 2-A. The electric power system. (*Source:* U.S. Office of Technology Assessment)

What are EMFs?

When electricity is moved long distances over transmission lines, leakage of electricity — called *corona* — occurs. (It has been estimated that some 40 percent of the electricity produced in the United States at any given time is wasted due to the corona effect.) Over the past ten or fifteen years, as new technologies have been developed, transmission lines that can handle higher and higher voltage have been used in order to reduce corona; the higher the voltage, the lower the leakage.

As the demand for electricity continues to grow, there has also been a steady increase in both the number and length of the power lines crossing the country. According to the Department of Energy (DOE), there are currently 350,000 miles of transmission line and some 2 million miles of distribution line in service in the United States.

Transmission lines end at regional substations, where the distribution components of the power system take over. Step-down transformers at the substations reduce the power to the lower-voltage — 115/230 V (wires with these voltages are run next to each other or in *bundles*), 30 amps — electricity that's used in the wall wiring of buildings and homes. (An ampere or amp is a unit used to measure electrical currents.) Then distribution lines, capable of handling up to 900 amperes of current, move the electricity on to the houses themselves.

There are two kinds of distribution lines or wires: primary and secondary. *Primary lines*, or *primaries*, are thinner than secondary lines. They're found at the top of electric poles, connected by porcelain insulators. (You can judge the amount of voltage in the wires attached to a pole by counting the insulators; generally, each one is attached to a line carrying 15,000 V, but this may vary from city to city.) *Secondary lines* are lower on the poles. They are distinguished by the amount of electricity they can handle. Primary lines carry 5 to 35 kV of electricity from the local step-down substations to small transformers that are either buried or mounted on wooden poles (referred to as pole-mounted transformers) near buildings.

Just go outside and look up at the power poles on your street — you'll probably spot a metal box similar to the one pictured in Figure 2-A. That's a pole-mounted step-down transformer. Its job is to reduce the voltage to 115/230 V so it can be used in the wall wiring in your home. Also note the thinner wire at the top of the pole that goes directly to the transformer and then to the house nearest it — that's a primary line.

Distribution lines take the 115/230-V current from the power pole right up to your door. Certain large appliances, such as your electric drier, run on 220 V of power; most small appliances use 110 V.

The electric company delivers power to your home or office at the *service drop*, the spot where your electric meter and fuse box are located. From

WARNING: the electricity around you may be hazardous to your health

there, the *customer end* of the system begins. A private electrician working for the builder installed the wall wires and outlets in your home, according to the municipal electrical code. Wall wiring consists of three lines: one energized (hot) wire carrying the full voltage; a neutral wire carrying return current; and a ground wire which should carry no current. These wires should, but don't have to, run alongside one another through the walls, thus allowing the neutral wire to "cancel out' the charges from the energized wires. (This setup also eliminates a magnetic field problem.) The National Electric Safety Code (NESC), however, has no provision to regulate the magnetic fields created by the arrangement of the wires in your walls. The code is only concerned with protecting you from shocks or fire that could result from the buildup of an electrical charge.

To prevent shocks or fire, the neutral and ground wires are joined together at the service entrance switch box and grounded. This means two things: Most important, a good connection must be made from the neutral line back to the transformer (this is the bare cable running to your house from the pole). This connection guarantees that your fuse or circuit breaker will trip if there is an electrical fault in your wiring. Secondly, a solid connection must be made to a metal electrode that is buried in the ground by your house. This helps maintain the electrical system you are exposed to at no more than about 120 V to ground and also plays a part in protection in the event of a lightning strike or freak accident to the power line. This second grounding connection is usually made to a rod driven into the ground and/or to a metal water pipe that runs underground.

The power you use at home is measured (on your meter and on your bill) in kilowatt-hours (kwh): 1 kwh is equal to 100 w of electricity used for 10 hours.

ELECTRIC CHARGES AND THEIR EFFECTS

No one has ever seen electricity, and we can only describe it by looking at the effects it produces. An *electric charge* is a basic property of certain elementary particles within an atom that cause them to attract or repel each other. There are two types of electric charges, negative and positive. The *electrons* in an atom carry a negative charge, while the atom's nucleus, its *protons*, carry an equal amount of positive charge. Opposite electric charges attract each other; charges of the same type repel each other.

The phenomenon of attraction actually binds electrons in an orbit around the nucleus of each atom, and, on a larger scale, binds atoms and molecules together to form the solids, liquids, and gases that make up our physical world. An ordinary atom — in fact, most things in nature — is electrically neutral because it contains an equal number of electrons and protons. Its overall charge is said to be zero.

What are EMFs?

But by rubbing materials together, we can upset this balance by altering the number of electrons in the atom — that is, either adding them or subtracting them. The materials are then said to be *charged*. In the physical world, electrically charged objects exert a force on other objects. It is this charge and the potential force it can exert that humans have been able to harness for electricity. A unit of this potential force — or potential electric energy — is called a *volt*.

The effect of these different charges on each other is what produces the voltage in an electric circuit. *Voltage* is the potential amount of electrical energy in a line — what people generally think of as the power. *Current*, on the other hand, is the actual flow or motion of the electrons or the electrical charge through the line.

To understand the difference between voltage and current, think of the water in a hose. The pressure of the water when it's not in use is the voltage. The current is akin to the water as it's moving through the hose when you're watering the lawn.

Electrical energy is measured in volts (V) or kilovolts (kV): 1 kV is equal to 1000 V. Current is measured in amperes (amps). The electrical power of a line is commonly measured in watts (W). The wattage of a line is obtained by multiplying the voltage by the current.

In North America, we have what is known as a 60 hertz (Hz) alternating current (AC) power system. *Alternating current* simply means that the electrical current that runs through our power lines and wall wiring doesn't just flow in one direction — it alternates back and forth. (The current in batteries flows steadily in a single direction, so it's called *direct current*, or DC.)

A *hertz* is a measure of alternating current. It indicates how many times per second a particular current cycles, or changes direction. For example, 1 Hz current will only change direction once every second; 60 Hz current reverses itself sixty times a second. (Europe runs on 50 Hz.)

Electricity, or electric power, travels along wires or *conductors* that allow electric charges to flow through them. (Some substances are better conductors of electricity than others. Good conductors — like copper, silver, or gold — contain a lot of electrons that have been dislodged from their orbits — so-called free electrons. Bad conductors — like rubber or glass — have few free electrons and are called *insulators*.) An electric circuit is a closed path made of conductors that the current of electrons can follow. For instance, when you turn on a lamp, the reason it lights up is that you've moved a band of metal that closed or completed the circuit from the outlet to the bulb.

The arrangement of the wires in the power lines themselves, as well as in wall wiring and appliances, is also known as a *circuit*. In this country, all

WARNING: the electricity around you may be hazardous to your health

the transmission and distribution lines have what is known as a three-phase power network, consisting of three hot or energized wires (wires with electricity flowing through them) in a bundle that work together. What you see as a single power line is actually two or three conductors that have been strung together. All three of these wires operate at 60 Hz AC, but they're arranged so that they're out of phase with one another, meaning they reach their peak voltages at different points in the 60-Hz cycle, generally one third of a cycle (or second) behind one another. For transmission lines, the combination of all the lines strung from the arm of the transmission tower is often called a circuit.

In a 60-Hz system, just as the direction of current goes back and forth sixty times a second, the voltage in the lines also shifts sixty times a second. The voltage reaches its peak, drops back down to zero, then climbs up to its peak again, sixty times each second.

Generally, the 115-230-V wall wiring in our homes and offices only makes use of the power in one of these three hot wires. Most factories, on the other hand, use all three of them for the extremely large loads of electrical power necessary to run heavy machinery, like the presses in an automobile assembly line.

It is possible to arrange the wires in power-line circuits and in appliances so that they cancel out or lower the magnetic fields they generate. This setup is used in many high-voltage transmission lines.

Electromagnetic fields (EMFs). Any time an electric current runs through a wire or an appliance, it produces an *electromagnetic field* (EMF). Electromagnetic fields are found wherever there's electricity and around any object that has an electric charge. *Electromagnetic radiation* is the wavelike fluctuation of the electric and magnetic fields that are produced when the electric charges are accelerated.

In today's electrical environment, EMFs are everywhere — near power generators and substations; under power lines; around radio and transmission towers; near ground currents in plumbing; and near electrical outlets, lights, appliances, office machines, and video terminals. In the case of our 60-Hz electric power system, and all the things that run on it, the fields it produces are called 60-Hz electromagnetic fields, which means the fields fluctuate in waves that go back and forth sixty times a second.

Besides man-made electromagnetic fields from our power system (sometimes called power-line fields), electric and magnetic fields occur in nature and in all living things. These natural fields are known to interact with man-made fields.

Electromagnetic fields (EMFs) is actually a blanket term for two different kinds of fields: electric fields (EFs) and magnetic fields (MFs) produced

What are EMFs?

simultaneously anywhere electricity is in use. Normally, electric fields and magnetic fields occur together. Taken together, they're commonly referred to as "electromagnetic fields." In certain instances, however, it's useful to discuss them as separate entities with different qualities and effects. This is especially true when we consider their health risks.

Electric field describes the electrical charge surrounding an object — that is, the region where the electric force of one charged object on another can be detected. An electric field is a measure of the electric force a charged object can exert on other objects entering its field. *Magnetic field* refers to an object's magnetic force, which is the force a moving electric charge produces on other moving charges. The same electrically charged object, say, a toaster, or a power line, will produce both an electric field and a magnetic field — that is, an electromagnetic field.

Besides being produced together, electric fields and magnetic fields have many common characteristics. They generate radiation in the form of waves. They're produced by electricity. They drop off as you move away from their source, diminishing according to the source itself. For example, fields from appliances disappear dramatically in a matter of a few feet, whereas a field caused by transmission lines may go on for hundreds of feet. Both fields are silent and invisible and, since people aren't really biologically equipped to consciously detect them, they largely go unnoticed even though we're surrounded by EMFs all the time. (A strong electric field will sometimes produce buzzing and crackling noises, or the sort of static electricity that makes the hair on your arms stand up. Extremely strong electric fields will cause what experts call a *contact current* — that is, a shock.)

But electric fields and magnetic fields are actually quite different in many important ways. In particular, they are known to have very different effects on living things. It's generally understood that when EMFs interact with people and animals, the electric fields and the magnetic fields separate, or *uncouple*, and affect the organisms separately. Whereas electric fields are present whether an appliance is on or off (as long as it's plugged in), a magnetic field will disappear as soon as an appliance is turned off. Whereas electric fields can be shielded by many things — houses, trees, and so on — magnetic fields are very difficult to shield because they're able to pass right through anything that doesn't contain a high degree of iron. This difference is crucial because it is believed that the dangers of electromagnetic fields come from their magnetic field component and not from exposure to the electric fields.

Electric field strength is measured in volts per meter (V/m). A kilovolt (kV) is one thousand volts; very high electric field strengths are measured

WARNING: the electricity around you may be hazardous to your health

in kilovolts per meter (kV/m). Weaker fields are measured in millivolts — one-thousandth of a volt — per meter (mV/m).

The strength of a magnetic field is measured in gauss (G) or in milligauss (mG). A milligauss (mG) is one-thousandth of a gauss. Most of the fields you'd expect to find around you will be measured in mG because field strengths measuring in gauss are extremely high. A typical home would have ambient fields somewhere between 0.5 and 4 mG. The international or industry measuring unit is the microtesla (μT); one microtesla equals ten mG; one tesla (T) equals 10,000 gauss. Microtesla is the unit you'll find often used in studies. Magnetic fields are measured with *GAUSSMETERS* or *MAGNETOMETERS*.

Electric fields are produced by the electric charges in power lines and lights and appliances, and their voltage remains constant. Magnetic fields result from the motion of these charges; that is, the current, and they fluctuate with that current. The stronger the current, the stronger the magnetic field. And just as the current varies according to the power usage, so do the magnitudes of the magnetic fields created by the current. (This is why, at peak usage times, magnetic fields around distribution lines are often as high as those found in the vicinity of high-voltage transmission lines.)

To calculate the magnetic field of a power line, 1 amp/m is equal to 12.6 mG field. For a 500-kV transmission line, the maximum magnetic field 1 meter above the ground, directly under the middle of the span (called the sagline), will be about 350 mG.

In our everyday life, we're involuntarily exposed to numerous *ambient* EMFs — that is, the ordinary fields produced by the power lines around us, as opposed to fields that are found only in specific workplaces or around a particular electrical device. Furthermore, a variety of ambient fields is quite likely to occur in any one place at any one time. For instance, think about the magnetic fields you might be exposed to while waiting for a bus. There may be fields from the electric streetcar system, fields from the power lines overhead, fields from a distant radio transmission tower, as well as the earth's natural geomagnetic field. And all these fields can, and do, interact— by adding to each other, subtracting from each other, at times even canceling each other out. It is this effect that makes the question of dosage in ambient exposures very complicated.

Numerous studies have established that most high magnetic fields in houses are produced by nearby power lines. Home appliance use has been found to be merely noise in the overall EMF picture in a home, although some particular household appliances — most notably electric blankets, hair dryers when used on children (effects on adults haven't been determined), black-and-white televisions, and certain kinds of electric heat —

What are EMFs?

have been linked to increased cancer rates. Your body itself has internal fields that are important to the regulation of particular cells. Some of these fields hold, or *bind*, molecules together; others play critical roles in the transmission of information from one cell to another.

On a grander, but not necessarily more important, scale, there are huge natural EMFs all over the earth and out in space. The earth has a natural 500 mG DC magnetic field that varies according to where you're located. It measures 350 mG at the equator, and about 670 mG at the magnetic poles. (It is this natural magnetic field that makes compasses work their magic.) Atmospheric processes, like lightning or sun spots, create changes in natural EMFs that can affect our electric power system. According to an article in *The New York Times* in June, 1991,

> A severe disturbance of the Earth's magnetic field caused by temporary changes in solar activity began on Tuesday night, threatening electric utility equipment and communication systems. . . . Utility managers around the country were notified yesterday to remain on alert, because such storms can interrupt electrical transmissions and damage transformers at generating stations. Some officials have questioned whether such a storm in April might have caused explosions and a fire at the Maine Yankee nuclear plant in Wiscasset.

The most important new finding about this DC natural magnetic field is the discovery that some particular strengths of the earth's geomagnetic field, in combination with AC magnetic fields over 2.5 mG, account for even greater increases in cancer than the power-line fields alone. In any case, you can see how difficult it would be, in today's electric environment, for any of us to escape exposure to electromagnetic fields. This is especially impossible for city dwellers.

Furthermore, not only are we being exposed to electromagnetic fields all the time without knowing it, but the strengths of the fields we experience are constantly changing as we go from building to building, or move through the rooms in our homes, or simply turn appliances on and off. We're actually creating our own little electromagnetic scenarios that bear a resemblance to the experiments going on in laboratories worldwide as scientists try to figure out what exactly EMFs do to biological systems and how they do it. In this case, the experimental animals are ourselves and our families.

The Electromagnetic Spectrum. To better understand electromagnetic fields, you'll need to have a sense of the whole range of electromagnetic

WARNING: the electricity around you may be hazardous to your health

energy or radiation as it exists in the universe. The *electromagnetic spectrum* (Figure 2-B) covers the wide range of electromagnetic radiation or energy. The spectrum is arranged according to both frequency and wavelength of the various kinds of radiation. Frequency is the number of waves per second that a particular form of energy sends out. *Wavelength* is the distance between two successive peaks of the wave. Higher-frequency radiation has a greater number of waves of shorter wavelength, and vice versa. Figures 2-C and 2-D give comparisons of the frequencies and wavelengths of various electromagnetic sources.

At the high end of the spectrum, with frequencies of billions of hertz, come the gamma rays, x-rays, and ultraviolet rays. These are known as *ionizing radiation*, because they have enough energy to enter cells and break chemical bonds, to get inside atoms and rearrange particles. Exposure to these waves can kill instantly, if exposure is high enough, or cause cancer.

All other forms of radiation are *nonionizing* — that is, they don't have the energy to actually get inside cells and break chemical bonds. It is this nonionizing radiation that we will discuss.

Middle frequencies come from visible light (10^{15} Hz), heat (or infrared radiation), radio frequency (RF), and radar and television waves known collectively as *microwaves*. Microwave radiation is known to be dangerous, because of its *thermal* effects — it can heat or cook tissue — and also because of its nonthermal effects, where certain biological changes occur but the body doesn't heat up. Microwave radiation is emitted by broadcast transmissions, radar, satellites, CB radios, RF sealers, electrical security systems, telephone relays, sonar, VDTs (video display terminals), microwave ovens and cell phones.

Extremely low-frequency (ELF) radiation is found around 60-Hz power lines and appliances. ELF radiation is nonionizing and nonthermal and, because it doesn't have enough energy to break up molecules or heat tissue, it was never thought to be dangerous. But very recently, scientists have discovered that the so-called "weak" ELF fields can indeed cause cancer and other diseases.

A HUMAN ANTENNA

By now, you're undoubtedly wondering, What exactly do electromagnetic fields do to us? The answer is both complex and disturbing. For one thing, we probably absorb a lot more energy from them than we realize. Furthermore, this energy has been found to have some measurable and potentially dangerous biological effects — *bioeffects* — on animal and human tissues and cells.

All matter is divided into *conductors*, which can transmit electricity by allowing it to flow through them, and *insulators*, which do not transmit

What are EMFs?

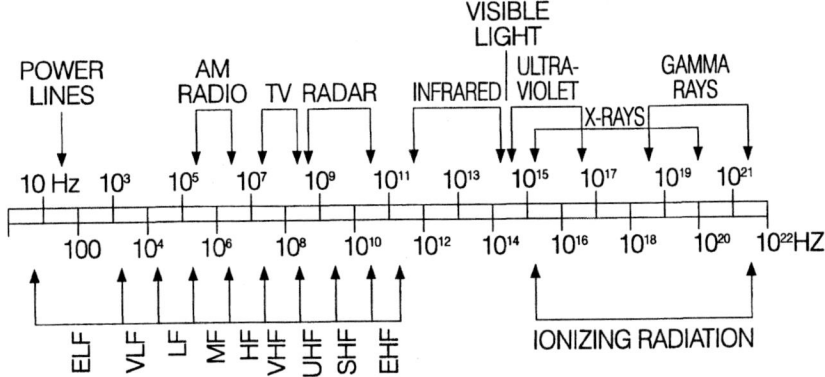

Figure 2-B. The electromagnetic spectrum. (*Source:* U.S. Environmental Protection Agency)

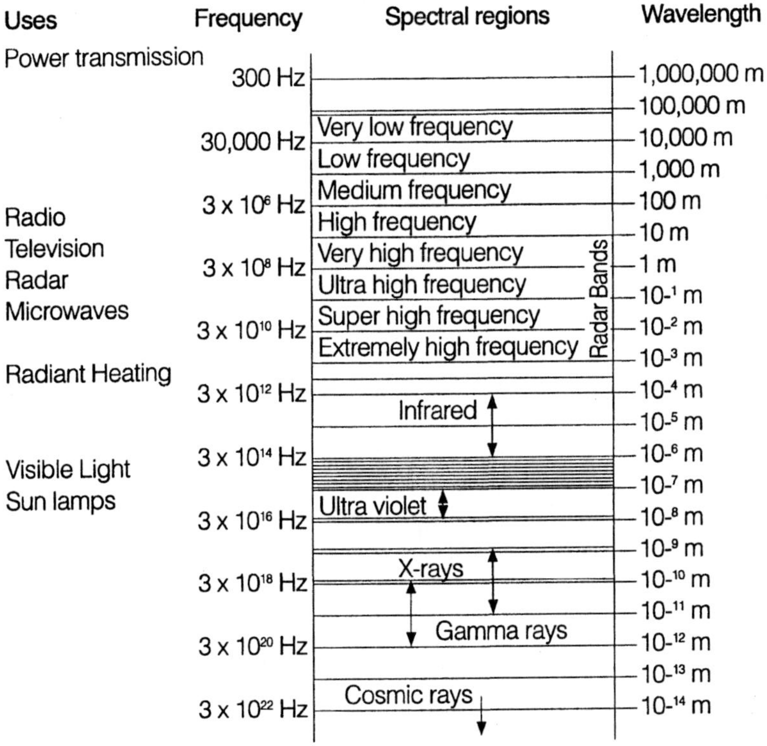

Uses	Frequency	Spectral regions	Wavelength
Power transmission	300 Hz		1,000,000 m
			100,000 m
	30,000 Hz	Very low frequency	10,000 m
		Low frequency	1,000 m
	3×10^6 Hz	Medium frequency	100 m
Radio		High frequency	10 m
Television	3×10^8 Hz	Very high frequency	1 m
Radar		Ultra high frequency	10^{-1} m
Microwaves	3×10^{10} Hz	Super high frequency	10^{-2} m
		Extremely high frequency	10^{-3} m
Radiant Heating	3×10^{12} Hz	Infrared	10^{-4} m
			10^{-5} m
	3×10^{14} Hz		10^{-6} m
Visible Light			10^{-7} m
Sun lamps	3×10^{16} Hz	Ultra violet	10^{-8} m
			10^{-9} m
	3×10^{18} Hz	X-rays	10^{-10} m
			10^{-11} m
	3×10^{20} Hz	Gamma rays	10^{-12} m
			10^{-13} m
	3×10^{22} Hz	Cosmic rays	10^{-14} m

Figure 2-C. The electromagnetic spectrum shown by frequency and wavelength. (*Source:* Bonneville Power Administration)

WARNING: the electricity around you may be hazardous to your health

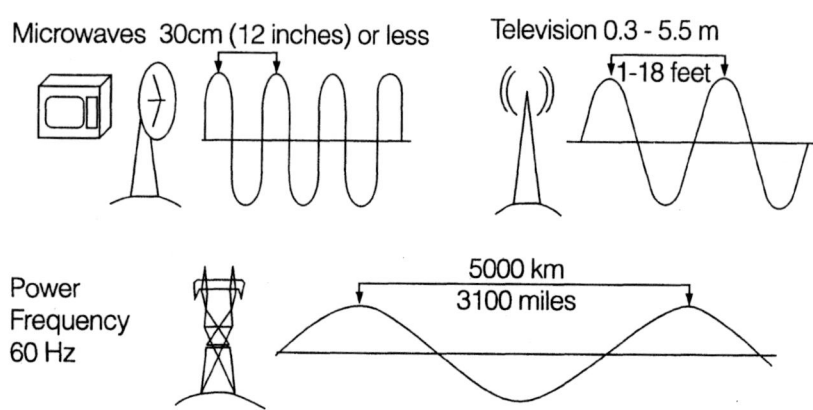

Figure 2-D. Comparison of frequency and wavelength for various electromagnetic sources. (*Source*: Bonneville Power Administration)

electricity. The human body, it turns out, is a very good conductor. Therefore, when you stand in an electromagnetic field, you become an antenna.

Furthermore, the human body has a higher conductivity than the air around it. But that's the good news, because it means an electric field will only produce some small surface currents on your skin and a limited amount of the electric radiation will actually enter your body. This amount will vary according to your size, shape or *geometry*, grounding, spatial orientation, as well as certain characteristics of the field itself. So, every time you're exposed to a 60-Hz AC electric field, 60 Hz currents will be set up in your body. For example, every time you touch an electric appliance, even though you can't feel it, little *contact currents* will run from the appliance into your body. Except in the case of a dangerous shock, these contact electric currents are of little concern.

The situation with magnetic field exposure is quite different. The human body has a magnetic permeability that is almost equal to air, so the entirety of a magnetic field will enter it. Even so, because humans can only detect magnetic fields in some very subtle ways, through the retina of the eye and the nervous system, you probably won't even notice the exposure. This, of course, is part of the problem. You may be sitting in the middle of a strong magnetic field and you won't even know it.

Chapter 3

Controversy and Cover-up

> *If I were a company, I'd have to protect my interests by hiring people who'd get up and say there is no problem. And some scientists never testify — they feel it either puts them on one side of the issue or diminishes their credibility.*
>
> JERRY PHILLIPS, M.D.,
> Jerry L. Pettis Memorial
> Veterans Hospital, Loma
> Linda, California

The notion that electromagnetic fields could be dangerous to your health is not an entirely new one. Warnings began as early as 1972, when scientists in the Soviet Union reported strange health effects in switchyard workers who were routinely exposed to high levels of electromagnetic fields. According to the Russians — to whom no one seemed to pay much attention at the time — the workers experienced increased heart disease, nervous disorders, blood pressure changes, as well as recurring headaches, fatigue, stress, and chronic depression.

As you can see in the EMF Timeline, the next fifteen years produced a slow, but steady, increase in information about the possible dangers of EMF exposure. It wasn't until the late eighties, however, with the publication of a number of key reports, that a red flag went up and people around the country started voicing their concerns and clamoring for some public recognition that EMF exposures were an actual health risk. In the late nineties, the controversy really heated up — with one side assuring the public there was no danger from EMFs and the other continuing to report a public health risk.

WARNING: the electricity around you may be hazardous to your health

EMF TIMELINE

1972 Reports from Soviet Union
1973 Navy Project Sanguine Committee
1975 Becker and Marino studies
1977 New York State PSC transmission-line hearings
1979 Wertheimer/Leeper study on childhood leukemia
1980 New York State Power Lines Project funded
1985 Texas school district wins first EMF lawsuit
1987 New York State Power Lines study reports
1988 Marcy-South class-action suit filed
1989 OTA report
1990 EPA report on EMFs as carcinogens
1991 Peters' report
1992 DOE funds RAPID
1992 California EMF effects project begins
1995 NCRP report is leaked
1995 Albom meta-analysis
1995 Kheifets EPRI meta-analysis
1996 NAS-NRC Review
1996 Miller meta-analysis
1996 Fear electrical workers' study
1996 Byus & Stuchly replicate mouse tumor study
1997 NCI Study
1997 Savitz meta-study of electrical workers
1997 Repacholi mouse tumor study (MW radiation)

1973: PROJECT SANGUINE

In 1973, a stellar committee of U.S. scientists attempted to issue a warning about disturbing biological effects of electromagnetic fields. The eight-member scientific panel had been commissioned by the U.S. Navy to review a series of studies on electromagnetic field bioeffects that were being conducted at the Naval Aerospace Medical Research Laboratories in Pensacola, Florida, in conjunction with the navy's proposed Project Sanguine in northern Michigan. When completed, Sanguine would involve 22,500 miles of underground cables that would produce an enormous electromagnetic field with the capability to beam radio frequency (RF) waves through the earth to establish communication links with the U.S. submarine fleet. The first phase of the project had already been started, but it had soon become a target of protest from locals who were worried about possible health risks from the fields.

The navy expected its inquiry to help defuse the opposition. Instead, the panel reported a number of worrisome findings — among them reports of birth defects in laboratory animals exposed to weak EMF fields, decreased task performance in exposed animals, and changes in the blood composition of human subjects after EMF exposure. (Even more alarming, scientists were finding similar alterations in blood chemistry in men working on the piece of Project Sanguine that had already been put in place.) Instead of smoothing the waters, the panel found the studies so alarming it recommended that the navy warn the administration about what appeared to be a potential major public health hazard from the 60-Hz electric power system. However, the navy ignored the recommendation, buried the report, and eventually aborted the project.

We can probably mark the year 1973, when the Project Sanguine advisory panel turned in that report, as the beginning of a trend on the part of the military, the power companies, and at times, even the government, to engage in a massive cover-up in order to keep people from learning that exposures to the ambient electric and magnetic fields produced by everyday 60-Hz power systems might be dangerous to their health. The reasons are obvious. The Pentagon itself produces a large majority of the electromagnetic fields affecting people today, in its weapons systems — which could be rendered obsolete if the magnetic field standards are revised downward — and radar towers. The power companies and large communication businesses have an economic interest in maintaining the status quo insofar as EMFs are concerned, because of the costs involved in reducing fields, because of issues of liability, and because of their strong commitment to "technological progress."

WARNING: the electricity around you may be hazardous to your health

The cover-up might have succeeded if it weren't for the actions of one of the scientists on the navy panel. Dr. Robert Becker, head of orthopedic surgery at the Veterans Administration Hospital in Syracuse, New York, and a known expert on bioelectricity, refused to be silent about what he believed was a serious threat to public health. Dr. Becker had been studying electromagnetic fields since the mid-sixties, in a series of experiments that began as an inquiry into the mechanisms by which electric currents promoted healing of bone fractures (the technique of applying electrical current directly to bone fractures has been used since the early fifties). He found that the electrical current promoted mitosis — that is, the current caused bones to heal by speeding up the process of cell reproduction. Early in his study, Dr. Becker had theorized that if electromagnetic fields could promote benign cell growth, they probably could promote malignant cell growth as well. This was later accepted as fact. Today, manufacturers of *bone-growth stimulators*, as the machines used in electrical bone healing are called, warn doctors not to use them on patients with known tumors. As Dr. Becker points out in his book, *The Body Electric* (Morrow, 1989), "Accelerated mitosis is a hallmark of malignancy as well as healing, and long-term exposure to extremely low-frequency (ELF) electromagnetic fields has been linked to increased rates of cancer in humans."

In further experiments, Dr. Becker and his colleague, Dr. Andrew Marino, exposed cancer cells to electromagnetic fields and reported they proliferated rapidly. (Other researchers would report similar findings in later studies.) In experiments at the Veterans Hospital, they found increased stress responses in humans and animals exposed to magnetic fields. Marino went on to do a generational rat study where he found that animals exposed to electromagnetic radiation had stunted growth, increased infant mortality, and changes in their blood composition and enzyme production.

Taken together, the experimental evidence they amassed convinced the two researchers that electromagnetic fields, such as the ones commonly associated with our 60-Hz power system, were dangerous and might pose a serious public health risk. Their findings prompted them to become two of the staunchest opponents of the unbridled growth of the electrical power system that was under way in the United States during the seventies and eighties. And they weren't alone. While Becker and Marino were conducting their experiments in New York, two other researchers, in California and Colorado, were quietly pursuing their own lines of questioning about the health effects of electromagnetic fields.

In California, Dr. Ross Adey, director of the Brain Research Institute at the University of California at Los Angeles, has been studying electromagnetic bioeffects for most of his professional life, having come to his electromagnetic field research by way of his research on the electrical charges that enable brain waves to pass through the cell membrane and carry messages to other cells. Up until the 1960s, it was generally accepted that only high-frequency radiation, such as gamma rays, x-rays, or microwaves, had biological effects. But, in laboratory experiments that went on for decades, Adey and colleagues Gavalas, Bawin, and Kaczmarek found that extremely low-frequency (ELF) electromagnetic fields — including electrical frequencies produced around power lines — had surprising bioeffects: Exposure caused changes in the behavior of cats and monkeys, altered their brain waves, and changed the level of calcium in chicks' brains. The intellectual excitement generated by this pioneering work is captured in an anecdote from Dr. Adey's 1991 acceptance speech when he was awarded the d'Arsonval Medal — the highest honor conferred by the Bioelectromagnetics Society — for his enormous contribution to the field:

> The excitement grew as we discussed a suitable tissue to test these RF fields for an effect on calcium binding. Bawin strongly favored the chick cerebrum, on which she was already conducting studies. I favored the rat, and Kaczmarek held his peace with typical British stoicism. . . . Like Professor Higgins, I am never one to take a position from which I will not budge, particularly with such a charming member of the opposite sex. And so the die was cast. We would use the chicks. But I exacted one condition. . . . We would test effects of a spectrum of ELF modulation frequencies, since my "female" intuitions were that we might find a frequency sensitivity at some point in the EEG spectral range. I shall never forget the excitement that mounted as, day by day, the spectral tuning curve unfolded, one frequency bar after another. . . . Finally, when the curve was complete as high as 35 Hz, a curious calm, even an unaccustomed silence, fell on the group. What to do next? There were needed repetitions before we could contemplate publication, but more than that, there was the humbling awareness that, if it were valid, here was an observation of a windowed response quite different than anything in the history of a biology based hitherto on equilibrium phenomena.

In Colorado, Nancy Wertheimer was an epidemiologist with the Colorado Department of Health. In 1976, acting without funding and with no clear notion of where the work would take her, Wertheimer began a study of children in Denver who had died of leukemia. Her study would become

WARNING: the electricity around you may be hazardous to your health

the bellwether for the notion that magnetic fields from the 60-Hz power system can cause childhood cancer. Wertheimer's is a classic story of a lone investigator who intuitively believes something is wrong, and, over great opposition, with few resources and little help, determines to go after that truth. Her methodology was simple enough: She obtained names and addresses of 344 children who had died of cancer, ran up a control group of another 344 children who didn't have leukemia, then got in her car and drove around to the deceased children's addresses to investigate. Initially, Wertheimer was looking for cancer clusters that might suggest some sort of viral agent, but she didn't find anything to support such a thesis. Being a trained observer, though, she did notice something else: A lot of the "death addresses" were close to pole-mounted electrical transformers.

At first that didn't mean anything to her, but then she happened upon a magazine article that said electromagnetic fields from power lines might cause cancer. She consulted with a friend, physicist Ed Leeper, and as they discussed her data, something clicked. From then on, Wertheimer began to focus on power-line electromagnetic fields instead of a virus, and her detective work paid off: She found that a significant number of the young leukemia victims had lived in houses where they'd been exposed to high magnetic fields in the last two years of their lives.

It wasn't until 1979 that Wertheimer and Leeper were ready to publish their results in the prestigious *American Journal of Epidemiology*, but word had already gotten around that a study in Denver had found children with high exposure to power-system electromagnetic fields were two to three times as likely to die of cancer than children with lower exposures. Overnight, Wertheimer was met with a barrage of protest and criticism, both from scientists who honestly didn't see how low-frequency radiation from power lines could cause any bioeffects at all, and from public utility officials with a clear interest in discrediting her work. Criticisms were leveled about her methodology — in particular, the fact that she had used a system of wiring code configurations to indicate the amount of current a line was carrying instead of measuring the actual fields.

From the start, Wertheimer said her work was only a beginning and called for additional research into what she felt was an important public health issue. However, at the time, no one seemed interested in replicating her study.

Backtracking to 1973, about the time the scientists on the Sanguine panel had made their recommendations, the New York State Power Authority was proposing to construct a 765-kV transmission line (the highest voltage technology permits) from Massena, near the Canadian border, to a small town named Marcy, 150 miles to the south. The line would bring hydroelectric power from Quebec to New York State consumers. Coincidentally, Dr. Robert Becker had a summer home in a little community right in the path of the proposed Marcy-South line. "Actually, it was just because I owned this piece of land up here that it happened," Dr. Becker recalls. "I happened to get a copy of the local paper and there was a notice in it about the power authority wanting to put in the line. I never thought I would be opening such a black box."

Dr. Becker opened the black box by leaking the Sanguine panel's final report to the New York State Public Service Commission, which had the power authority proposal before it. He also wrote a letter telling of the panel's wish to warn the public via the White House and explained that he and the other Sanguine scientists believed electromagnetic fields generated by ambient 60-Hz power lines all over the country were a serious public health hazard.

"If I hadn't opened up the environmental issue on this, I wouldn't be here now," Dr. Becker said recently, indicating his rustic homestead high up in the Adirondack Mountains. "I would probably still be at Upstate (Syracuse). But I don't think I'd want to change anything I've done, looking back." Then he explained his decision to bring the danger of electromagnetic fields to the attention of the public:

> You see, it wasn't a military secret. The report was never classified. I was very naive at the time. I thought I'd just give them the information and go home and they'd fix it. Well, I learned a lot from the experience. For one thing, I learned that scientists are just like anyone else — they can be bought.... The Department of Defense, the secret services, they're all involved in this. They've known for some time there's a real civilian problem and they've kept it covered up. You see, once you raise the environmental issue, once that dam is broken, reporters start scurrying around the edges and that means trouble. The military knows that.

The New York State Public Services Commission didn't hold public hearings on the proposed line until the late seventies, but by that time the EMF issue had heated up enough for the commission to ask Drs. Becker and Marino to testify. The scientists recommended that the commission disallow the line, citing their and other studies that had found dangerous biological effects from EMF exposure.

On the other side, however, the power authority was able to mount a strong defense, using paid "experts" who were completely without experience in EMF research, but were nonetheless eager to assure the public there was no cause for concern. (Three of these scientists were on the staff of the National Institute of Health, Drs. Aaronson, Sinks, and Tucker. They received a total of $125,000 from the power authority for testifying at the hearing and were eventually reprimanded by their director for ignoring agency rules about outside compensation for consulting. At the time, Dr. Adey questioned whether the agency would be impartial in future EMF research when their scientists had received so much money from a utility.)

The power authority was able to get Governor Hugh Carey to disregard the commission's concerns and, late in 1977, a permit for the Marcy-South line was issued. Once again, politics prevailed over public health. The controversy surrounding the project, however, was to result in the largest research study to date about the bioeffects of electromagnetic fields: the $5 million, five-year New York State Power Lines Project Study.

1980: THE NEW YORK STATE POWER LINES PROJECT

Luckily, the power authority wasn't able to stifle the commission entirely, and it attached several requirements to the permit for the line because of members' concerns about the possible dangers of EMFs. The commission went on record with a proviso to the permit warning that "Electromagnetic fields had been shown to produce bioeffects on animals and it is possible they might do the same in humans." A second provision established a 350-foot right-of-way along the power line route. Finally, the commissioners required the power authority, in conjunction with the New York State Department of Health, to finance a major research study to find out whether there were human health risks from electromagnetic fields produced by overhead power lines.

The Power Lines Project got off the ground in 1980, when a body of scientists and engineers was chosen for the Scientific Advisory Panel. The panel was selected on the basis of both professional expertise and lack of financial or professional conflicts of interest. In turn, the panel awarded contracts to scientists all over the country to carry out a total of sixteen studies on the biological effects of electromagnetic fields.

A $355,905 study, "Childhood Cancer and Electromagnetic Fields," conducted by Dr. David Savitz, an epidemiologist at the University of Colorado Medical School, turned out to be a milestone. Savitz was charged with replicating Wertheimer's "flawed" study, and it was well known that everyone expected him to discredit her. A lot of people were very surprised when Savitz supported Wertheimer's findings. To the scientific community and the utilities, the Savitz study was a bombshell.

According to the Scientific Advisory Panel's final report, "Biological Effects of Power Line Fields," released in July, 1987,

> Several areas of potential concern for public health have been identified. . . . Of particular concern is the demonstration of a possible association of residential magnetic fields with incidence of certain childhood cancers. . . . A more serious concern comes from a study of cancer in children suggesting that children with leukemia and brain cancer are more likely to live in homes where there are elevated 60-Hz magnetic field levels than are children who do not have cancer.

The Savitz study found a positive association between wiring configuration and increased cancer risk, just as Wertheimer had. This held for all childhood cancers, especially leukemias and brain tumors. There even appeared to be a dose-response relationship, something that had been missing from the earlier work. Savitz estimated that as much as 15 percent of the childhood cancer in the United States is caused by EMFs from power lines. Though the Savitz study reported a slightly lower risk ratio — one and a half times — than the Wertheimer study, his findings were taken much more seriously than Wertheimer's had ever been.

Another experiment on the clonogenicity (reproduction) of tumor cells by two scientists, Wendell Winters and Jerry Phillips, proved most interesting. Their findings came about almost by accident, and the final report of the Scientific Advisory Panel is careful to state that they were not part of the official Winters research project studying the effects of 60-Hz EMFs on human and canine cells. Dr. Winters had a state-of-the-art magnetic field laboratory in which to do his research. Dr. Phillips was working in another lab — not under the auspices of the Power Lines Project — studying cancer cells. The scientists got an idea to try something: "We were growing cancer cells in my lab," Dr. Phillips explains, "and we brought them over to Wendell's lab to expose them to EMFs. Then we brought them back to my lab to see what had happened."

When they exposed Dr. Phillips' cancer cells (they were human colon carcinoma cells) to Dr. Winters' magnetic fields, they proliferated like crazy. Furthermore, exposed cells became increasingly resistant to the body's immune system cells. These cells that normally fight tumors (natural killer cells) exhibited both structural and chemical changes. Drs. Phillips and

Winters stated that their observations led them to believe that it was possible that magnetic fields stimulate the rate of cancer cell growth, or act as a cancer promoter.

Whereas the Savitz study had been met with consternation, the Winters and Phillips study was met with derision. It took nearly a decade of EMF research — with Dr. Phillips at the Cancer Therapy and Research Center in San Antonio and Dr. Winters at the University of Texas Health Sciences Center — for that early experiment and others like it to be taken seriously.

Winters recalls, "When we first reported bioeffects, they challenged us as scientists and as people. Over the years, the bioeffects of magnetic fields have become a fact. Today it's accepted. There's no question now that magnetic fields have bioeffects. Across the spectrum, in all species and animals, there are responses to power-line frequencies."

The Power Lines Project was the first in a series of major studies to alter scientific perceptions about EMF risks. The effects are most clearly seen in the experience of the project administrator, Dr. David O. Carpenter, then director of the Wadsworth Center for Laboratories and Research of the New York State Department of Health. Today, Dr. Carpenter is the dean of the State of New York (SUNY) School of Public Health in Albany, New York. In the eighties, Carpenter was, for want of a better description, radicalized by Savitz' findings:

> At the beginning of the project, I did not think there was anything to it. I thought the whole area seemed like one filled with kooks and charlatans. In fact, everyone on our expert panel was equally skeptical. It didn't seem possible that 60-Hz power, which is of much lower frequency than visible light, could cause any harm. But our scientific conclusions changed my mind. No scientist can have that kind of experience without changing his mind.

Another thing to remember, Carpenter says, "was that these studies only dealt with the child's exposures when he (the child) was at home. All the other possible sources of exposure were not taken into account, so the danger is probably grossly underestimated." Carpenter continues:

> I am now convinced that EMFs pose a health hazard. There is a statistical association between magnetic fields and cancer that goes beyond the shadow of reasonable doubt. I think there is clear evidence that exposure to EMFs increases the risk for cancer. This is most clear with leukemia and brain tumors, but in

the residential studies, statistical significance increased for all kinds of cancer. And we're just beginning to have a whole body of evidence that reproductive cancers are increased by exposure.

And you have to remember, it's neighborhood distribution lines that are the concern. The (Savitz) investigation said it was likely that 10 to 15 percent of all childhood cancer came from exposure to distribution lines. Everyone worries about high-voltage transmission lines, but the study was about neighborhood distribution lines. What most people don't realize is current in a high-voltage line is often no higher than in neighborhood lines and the EMFs are about the same. There's a 50- to 100-mG field under neighborhood distribution lines.

For years, Dr. Carpenter has been one of the most vocal advocates of immediate and strong regulatory action to protect people from exposure to magnetic fields:

We need to really raise some red flags on this. I believe we had enough information three years ago to make changes. It's time to stop pussyfooting around. The evidence is very good right now. Some forty studies of electrical workers (see Appendix A for a review of the studies) show great increases in deaths from leukemias and brain tumors. That supports the childhood studies. We public health professionals should be telling people there are a few things you can do to reduce exposure and reduce the chance of you and your family getting cancer.

One of the main issues Dr. Carpenter is grappling with is: "How do we communicate problems where there is controversy in the scientific community, but where the public health community feels there's an exploding public health problem?" From the moment the Scientific Advisory Panel published its report, it should have become impossible for any conscionable person in the field to deny the health risks of electromagnetic fields. Yet, many continued to do so. In Dr. Carpenter's words, "One of the major problems here is that the few people who are highly visible on this have conflicts of interest. I think that everyone who has a thing to say publicly about the dangers of EMFs should have his income sources checked. And we need to make sure studies are financed by individuals who do not have any financial interest in the outcomes."

WARNING: the electricity around you may be hazardous to your health

By the early eighties, despite the face they were showing to the public, the power companies themselves had begun to worry about possible health risks of electromagnetic fields created by their lines. Here and there around the country, utilities had started to fund their own research — studies about measuring fields, bioeffects, even mitigation studies designed to develop technologies to reduce the magnetic fields coming from their lines. Unfortunately, many of the positive studies — experiments that prove the relationship between EMFs and bioeffects — never saw the light of day. In this way, the people controlling the purse strings have also controlled the flow of information.

One EPA (Environmental Protection Agency) scientist explains he's very disappointed with what's been going on:

> With the cutback in federal funding, researchers are being funded by vested interest groups, like EPRI (Electric Power Research Institute). These researchers are more inclined, because of subtle pressure, to only represent negative results when it comes to EMFs or not to report positive results. They're afraid to get painted into a corner as being environmental activists and then they'd have difficulty getting funded. That's why discussion in scientific circles is so muted.

The question of funding is crucial. On the one hand, there's a great hue and cry for further research — often used as a delaying tactic, or to suggest that there isn't enough information about EMF exposure to take any action. On the other hand, there's a dearth of government money available for EMF research. Agencies like the EPA that once had such research programs had them closed down as funds were withdrawn during the Reagan administration. The majority of current EMF research is under the auspices of the utility companies or the Department of Energy — hardly unbiased funding sources.

According to Andrew Marino, who was the editor of the journal *Bioelectromagnetism* for ten years,

> The question of funding EMF studies in the United States today is critical. Virtually all the money comes from sources that want negative information (proving the dangers of EMFs) covered up. That serves to put a cap on the research. You give me enough money, I can fill a courtroom with so-called experts who'll say whatever I want the jury to hear. There are a litany of well-crafted arguments against the positive research. Sophistry. Equivocal use of the word cause. Badly obscured data, all used to obfuscate and confuse.

Having seen the handwriting on the wall, the utilities started to surreptitiously make changes in the circuitry or siting of transformers and power lines to reduce the levels of public exposure. Citizen activists have reported instances of utility workers sporting gaussmeters who appeared around a questionable neighborhood electric facility, worked for a few days, then disappeared. Afterward, television reception — poor TV reception is a common marker for high magnetic fields — would suddenly improve or people would finally find relief from the recurring headaches and other physical problems they'd been experiencing for years.

1985: EMF LITIGATION

Court cases around the country also had the utilities' attention. One in particular, in Texas, was to cause them a great deal of consternation. The utility official who decided to file that suit is probably still reeling from the outcome of the case. In 1985, Houston Lighting and Power sued the Klein Independent School District because the school board had refused to grant the utility a right-of-way for a 345-kV transmission line that would run close to three schools.

The school district's attorney, H. Dickson Montague, was able to muster enough credible scientific evidence about the dangers of power-line fields to not only convince a jury to direct the power company to remove the line but to require it to pay the school district $104,000 in compensatory damages and an addition $25 million in punitive damages for its irresponsible behavior in siting the line so close to children's facilities. (The utility appealed and damages were reduced to $140,000.)

According to the appeals court, "At issue in this case is a relatively new scientific concern — the possible health effects or risks associated with exposure to high-voltage power lines. A related legal issue concerns the forum that should consider those effects." Three EMF researchers testified on behalf of the school district. Nancy Wertheimer told the jury that the children in each of the schools would probably experience an increased risk of cancer and that it was "indefensible" to expose them to that risk. Dr. Jerry Phillips testified that his laboratory studies indicated that EMF exposure causes cancer cells to grow faster than nonexposed cells and to become more resistant to immune system destruction. And Dr. Harris Busch, chairperson of the Department of Pharmacology at Baylor College of Medicine, said the energizing of the line had been an "inadvertent prospective experiment" about EMF effects on the health of the children. According to Montague, the case would have tremendous repercussions:

I believe the utilities are in serious trouble. We proved there was a potential for personal injury from power lines. There's a burgeoning area of EMF litigation now; more and more people want to file personal injury suits. Although, as you might expect, it is extremely expensive to fund a case such as this. It is fortunate that Klein Independent School District had the financial wherewithal to stand up to the utility company and present a full and complete picture to the jury on the potential adverse health consequences of prolonged exposure to electromagnetic fields. It is rare that a person is given this opportunity. However, when one is given this opportunity, the results are obvious.

Three years later, 143 property owners filed a pair of class-action suits against the New York State Power Authority, claiming the Marcy-South line had destroyed the value of their property by setting up a "cancer corridor" along its path (see Chapter 1). And a year later, in Seattle, Washington, Robert Strom, a longtime employee of the Boeing Co., filed a lawsuit that claimed his leukemia was due to on-the-job electromagnetic radiation. (The Strom case was settled for $500,000 in the fall of 1990, with the added stipulation that Boeing had to provide a medical program for its workers and monitor the health effects of EMF exposure.)

1989: OTA REPORT

In 1989, the Office of Technology Assessment (OTA) commissioned a study on "Biological Effects of Power Frequency Electric and Magnetic Fields" (to use as a background paper for OTA's assessment of "Electric Power Wheeling and Dealing"). The authors of the 102-page report were Granger Morgan and Indira Nair, two highly respected physicists in the Department of Engineering and Public Policy of Carnegie Mellon University, and H. Keith Florig, a researcher in the department. After reviewing the existing EMF research, the authors concluded, "The quality of the science that is now available is remarkably high" and "The emerging evidence *no longer allows* (emphasis added) one to categorically assert that there are no risks."

The OTA report should have put the utilities on notice that a big change in the tide of public opinion was imminent and that they couldn't continue to do business as usual. In fact, the power companies continued their conservative policies, insisting there was nothing about their power lines for the public to worry about.

Another important milestone occurred in 1989, when epidemiologist Genevieve Matanowski from Johns Hopkins University reported on a four-year study of 50,000 New York Telephone Company employees with varying EMF exposures. The study found workers with high exposures had higher rates of cancer — for some as high as seven times the expected rate — particularly leukemia and lymphoma. Perhaps the most important finding of the Matanowski study was that of a dose-response relationship: Workers with the highest exposure had nearly twice the amount of cancer of any other group.

That same year, Savitz and Loomis reported on a major study of occupational mortality from sixteen states showing that workers in the electrical professions had a much higher risk of brain cancer. (Back in 1985, Savitz had reviewed eleven occupational studies and reported one and one-half times the risk of leukemia for workers with EMF exposures, with evidence that linked the amount of exposure to the incidence of disease.)

But 1990 was to be the real turning point, the year when the public finally began to realize that a critical mass of experts was warning that electromagnetic fields were dangerous to their health. And it began to dawn on people that the danger came not from some far-off source but from the electric power lines all around them. A comprehensive *New Yorker* article about EMFs by investigative reporter Paul Brodeur, published in the summer of 1990, was an eye-opener for many readers. Brodeur's work broke the ground for a spate of articles in newspapers and periodicals and the occasional television show. By that time, studies were being reported that made the subject of EMFs — and their possible health risks — news.

1990: THE EPA REPORT

Perhaps the single most devastating development, from the point of view of those who were still trying to persuade people they had nothing to fear from electromagnetic field exposure, came in 1990, when the Environmental Protection Agency decided to embark on a comprehensive review of the available EMF literature. (This investigation was actually prompted by pressure from Congress to find out whether RF and microwave radiation were dangers to the public. EPA scientists decided to include research on ELF radiation in their review.)

The EPA is the federal agency charged with warning the public about health problems in the environment. Upon completion of its "Evaluation of the Potential Carcinogenicity of Electromagnetic Fields," a 150-page review of the available EMF studies, concerned EPA staff members tried to warn the people. But their efforts were blocked by a concerted group of detractors that went all the way up to the White House itself.

In December, 1990, frustrated by what he viewed as an attempt on the part of the administration to keep the report from the public, David Bayliss, a veteran EPA staff scientist who was one of the authors of the EMF review, leaked a draft of it to the press. Bayliss told reporters that the Pentagon and the White House had interfered with the document twice in the past year; once when science advisor F. Allan Bromley forced the agency to delete an early designation of EMFs as a class B1 carcinogen (a probable cause of cancer in humans — cigarettes are a class B1 carcinogen); then, again, when Bromley took the report, ostensibly for review, and kept it under wraps for nearly six months (Bromley has defended his actions by explaining that he "didn't want to alarm people"). Bayliss and other staff scientists felt their warning was never going to be made public.

Others felt the same. In an article in the prestigious British medical journal Lancet (vol. 337, March 2, 1991, p. 544), author J. B. Sibbison wrote:

> A major objection was that the report "might be unnecessarily alarming to the public." That is a public relations, not a scientific, problem. The U.S. Air Force, for one, must deal with public opposition to nonionizing radiation from the emergency communications network it is building. "If published, the (EPA) report will contribute to public anxiety and have serious impacts on capabilities and costs of air force programs," says an air force report. The power industry has similar public relations problems.

If it hadn't been for Bayliss, the public might never have learned of the EPA's concerns about electromagnetic fields. Bayliss, who was widely quoted in the press, made public both the results of the largest review to date on EMF research and the ongoing federal cover-up of the issue. That cover-up, in fact, is still very much in effect, for the agency has never officially released the report in its final form.

Dr. Doreen Hill, of the U.S. EPA EMF Group, and co-author of the draft review, said later,

> In one sense, the report is out to the public anyway. This has turned out to be quite a political and scientific controversy. One good thing about the controversy surrounding the report is that finally the subject has gotten someone's attention. Now, maybe this will get the research funding it deserves. When you get these high-level people involved, things are bound to start to happen. Before this, EMFs were treated as a non-issue.

The EPA reviewed studies that had been completed prior to 1990 that examined the correlation between EMF exposure and cancer. These included six residential studies of children and adults, over thirty studies of workers in electrical occupations, two studies of the relationship between children's cancer rates and their fathers' EMF exposures, as well as hundreds of laboratory studies. In part, the review found:

> The several studies showing leukemia, lymphoma, and cancer of the nervous system in children exposed to magnetic fields from residential 60-Hz electrical power distribution systems, supported by similar findings in adults in several occupational studies also involving electrical power frequency exposures, show a consistent pattern of response that suggests, but does not prove, a causal link.... Evidence from a large number of biological test systems shows that these fields induce biological effects that are consistent with several possible mechanisms of carcinogenesis.
> The strongest evidence that there is a causal relationship between certain forms of cancer, namely leukemia, cancer of the nervous system, and, to a lesser extent, lymphoma, and exposure to magnetic fields comes from the childhood cancer studies. ... In two of these studies, cases were observed in those exposed above 2-3 mG, but not in people exposed below that level.

Using what the authors describe as an "overall weight of evidence approach," the review concluded, "The evidence for a causal relationship (between EMFs and childhood cancer) is too strong to dismiss as chance and not strong enough to be regarded as proof of causality."

The national press was united in reporting that the EPA felt that electromagnetic fields cause cancer. *The Boston Globe* stated (January 14, 1991):

> A federal survey that calls magnetic fields from electric power lines a "possible" cause of cancer has given heightened urgency to a growing national effort to settle the issue. . . . The EPA concluded that several studies "show aconsistent pattern of response that suggests a causal link" between power lines around homes and cancer. The draft said flatly that "there is a link between EM fields and certain forms of site-specific cancer."

Paul Raeburn reported (Associated Press, Washington, D.C., January 14, 1991):

An Environmental Protection Agency report linking electromagnetic fields to leukemia and brain cancer in children will be released next week after being held up by the White House science advisor.

Time magazine said (December 24, 1990, p. 67):

> The EPA has put forward what amounts to the most serious government warning to date. The agency tentatively concluded that scientific evidence "suggests a causal link" between extremely low-frequency electromagnetic fields and leukemia, lymphoma, and brain cancer . . . (the report) does identify the common 60-Hz magnetic fields as "a possible, but not proven, cause of cancer in humans."

Bayliss, who has been reprimanded for talking to the press, explained,

> You should know that some of the scientists here at the agency don't agree with the way the administration's handling this. We're under pressure from the White House. You have to realize, the power companies are some of the biggest backers of this administration. And the air force is building a string of radars across the country, so they're afraid they'll be stopped. They all must feel their economics are being threatened.

Bayliss says other scientists at the agency share his concerns about the apparent EMF health risks:

> I'm particularly concerned with the epidemiological data that show an elevated risk for children who are exposed to EMFs in excess of 2 mG. Even though we don't have conclusive causal data, in my opinion the thrust of the data warrants concern. The public must be made aware there is a risk involved in exposure to EMFs.

What sort of a risk are we talking about? Critics of the review say that the identified risks aren't large enough, statistically, to warrant concern. They also suggest that misclassification of actual exposures may have "biased" the results. But, the review points out that random exposure misclassification tends to bias relative risks toward the null, meaning that with more precise measurements, the risks would actually be *higher*.

Controversy and Cover-up

Bayliss explained,

> When we only see a slight excess, that doesn't mean there's no risk. It means we haven't yet been able to isolate the risk. Right now, we see a slightly elevated risk in the cases of childhood cancer. I think it's irresponsible for people to go around saying it's safe. To me, it's not a public health position at this point not to urge caution.

Speaking of exposure measurements, Dr. Carpenter reminds us, "All these studies have ignored the child's experience when he's not at home, so we're probably grossly underestimating the danger."

That was hardly the end of the battle over the EPA report, however. Since then, it's been subject to continuous review, by committee after committee, as the government and the power companies delay its release or try to bury it. In the spring of 1992, when a scientific review panel was being named, Bromley tried to stack the deck by recommending a majority of conservatives and scientists who had historically testified on the utilities' side of the issue. After a struggle, a neutral review board was named. In July, the panel of experts concluded that the report was excellent. But, in August, at a White House Appointed Science Advisory Board meeting in Texas, with Electric Power Research Institute (EPRI) and conservative members of the advisory board dominating the meeting, the report was denounced as "biased," and its authors were directed to rewrite it. At the meeting, attention centered around a fifty-page review, by Dr. George Hutchison, that had been commissioned by EPRI. Dr. Hutchison concluded there was no evidence that EMF exposure was connected with cancer.

"I find it amazing that Bates (David Bates from Vancouver) and Clark Heath, who never did anything in the EMF field, and Pat Buffer, who's on the EPRI payroll, accuse us of bias!" says Bayliss. "They had direction all the way up to the White House, from F. Allen Bromley. They've been told by the White House to tone it down." Prior to the San Antonio meeting, the EPA staff had no knowledge of the Hutchison paper. Bayliss explains:

> What we finally found out had happened was that EPRI paid Hutchison to write this review. We never even got it until after the meeting. It's very difficult to review the studies and deny the connection — there's so much evidence. We did get a copy. He treated all the rotten negative studies as valid. A lot of industry-sponsored studies are done so that they'll turn out negative, then they're used to evidence that there are no effects. We

hired Charlie Poole (of Epidemiological Resources of Massachusetts) to review Hutchison's paper. He was quite critical of what Hutchison did.

Dr. Poole said he questions the usefulness of the type of statistical analysis Hutchison used. "My view of that paper is that it was mostly a statistical exercise. He aggregated all the studies and came up with a finding."

According to Martin Halper, director of the Radiation Studies Division of the agency, the advisory board report actually strengthened the EPA document, stating: "A weakness often found in such studies is confounders. But since no common confounders can be identified, the existing evidence cannot be dismissed." (A confounder is an extraneous factor that distorts the true relationship of the variables being examined in a study.)

Bayliss thinks it's time for the cover-up to end:

> They've got to get rid of the connection between the White House and the regulatory agencies. We sit here now and we work on this stuff and we realize it's not going to go anywhere. In this group, right now, practically all the staff scientists are Ph.D.s. We used to sit on the carcinogen board, but all that power's been taken away from us. It's the political appointees, many of them just B.A.s with no scientific background, who are now speaking for the agency. I'm really very disappointed in what's going on.
>
> At the very minimum, people in this country should be told there may be a danger living in magnetic fields. And that caution should be exercised. Utilities should remove the fields from around children. The EPA should be on the forefront of warning that any community wishing to build a school should not build it under overhead lines and should be careful to measure the fields first.

But that is far from what's happening. The authors were instructed to revise the document and have it reviewed by "all interested parties." "If we have to get the power companies to agree, it'll amount to a whitewash of the subject," Bayliss contends. "They'll move heaven and hell to get us to alter what we've said. I've seen it before. It happens any time you say a product is dangerous."

When the revised report finally went out in 1994, it was entitled, "The Relationship Between Power Frequency Electric and Magnetic Field Exposures and Human Cancer," and the authors had dropped the earlier recommendation that power-line EMFs be classified as a carcinogen. But the authors still clearly believed there was a real association between EMFs and cancer. The report emphasized that new, stronger studies linking childhood cancer to EMF exposures were appearing all the time.

But, while the EPA controversy was still brewing, something unexpected happened: A utility-funded study took the existing EMF cancer data a giant step forward.

1991: JOHN PETERS

The key study everyone in the scientific community was waiting for in the nineties had been funded in 1986 by the Electric Power Research Institute. It was a large-scale, epidemiological study of childhood leukemia and electromagnetic fields that was being conducted by Dr. John Peters of the University of Southern California (USC) School of Preventive Medicine. The Peters study was expected to take the Wertheimer and Savitz data a giant step forward because of the size of the study population, its careful design, and the fact that the researchers took twenty-four-hour field measurements in the children's homes. (It turned out that Wertheimer's wiring code configurations were still a better indicator of the risk.)

Early in 1991, not long before Dr. Peters was due to give a preliminary report at a closed-door EPRI session in Carmel, California, word on the street had it that he was going to replicate the earlier studies. The news caused quite a stir in the scientific community: Despite his power company connection, Dr. Peters was known to be a careful scientist and a man of great integrity. (A story about him had made the rounds earlier that year. At an EPRI meeting in Texas, everyone had been polled as to whether or not they would buy a new home near a high-voltage transmission line. Dr. Peters was the only one in the room to say "no." This had been preceded by an incident at a 1986 conference in Canada, where Richard Phillips, an EPA scientist, announced to a roomful of people that he'd never buy a home near a high-voltage right-of-way.)

At the Carmel meeting, Peters did exactly what he'd been expected to do: He reported that exposure to high magnetic fields increased childhood leukemia risk up to two and a half times. He also reported that children who frequently used hair dryers and watched black-and-white television had double the risk of the disease.

EPRI is the largest conglomerate of utility companies in the world. Its headquarters is in Palo Alto, California, and it has the financial wherewithal to endow an enormous amount of EMF research. But sometimes EPRI doesn't like the results of the studies it funds, and it also has a history of spending a lot of money obfuscating the EMF health risk issue by publishing misleading claims and withholding potentially damaging data. Dr.

Carpenter and other scientists have gone on record repeatedly stating that EPRI should not be allowed to finance or report on any EMF research.

At the Carmel meeting, EPRI was true to its colors, barring the press from the meeting and preventing Dr. Peters from granting interviews. The "gag order" on Peters held until November, 1991, when his report came out in the *Journal of Epidemiology*. Dr. Peters' reluctance to discuss the subject can perhaps be better understood in light of the bombshell that he dropped in early November at the annual Department of Energy (DOE) review on nonionizing radiation.

Dr. Peters and Dr. Joseph D. Bowman of the National Institute for Occupational Safety and Health (NIOSH) had done a separate analysis of Peters' data and made an exciting discovery. The risk of childhood leukemias increased from exposure to certain combinations of the earth's DC magnetic field and manmade 60-Hz AC magnetic fields. By combining the two measurements, Drs. Bowman and Peters had obtained a dose-response curve for leukemia risks. The risks they reported when static and AC field measurements were combined were far greater than any that had ever been obtained before: The increased risk for childhood leukemia went from the previously noted two or two and a half times to six and even nine times. This was the first time a dose-response relationship had been established for EMF exposure and cancer.

Interestingly, none of this appeared in the Peters report that was published that November, which omitted any mention of the earth's magnetic field measurements. It's not difficult to speculate that this had something to do with EPRI, which, from the start, has tried to downplay the results of the study. Right after the spring meeting in Carmel, when Peters released his preliminary findings, EPRI had distributed a conservatively worded press packet. But that didn't prevent the press from sounding an alarm, as headlines like the following appeared the next day all over the country: STUDY SHOWS MAGNETIC FIELD EXPOSURE MAY INCREASE RISK OF LEUKEMIA; MAGNETIC FIELDS CAUSE LEUKEMIA; MAGNETIC FIELDS RESULT IN TWO AND A HALF TIMES THE CHILDHOOD CANCER; LEUKEMIA AND HOME APPLIANCES.

The press packet contained a summary entitled "EPRI Commentary on Initial Results from the USC Study of Childhood Leukemia and Exposure to Electric and Magnetic Fields." That paper provides an excellent model of the techniques that vested interest groups employ to do damage control on a potential "hot potato" like this one.

The opening paragraph warns that "interim results" given in the Carmel workshop "should be considered preliminary and are subject to revision." A common means of misrepresenting scientific data is to reorganize and restructure it statistically until it comes out the desired way.

Throughout the paper, negative findings — where no correlation was found between EMF exposure and cancer — are given emphasis, while positive findings — where the link between EMFs and cancer is shown — are downplayed. The strongest finding of all and the most potentially damaging to the utility industry — the finding that establishes a relationship between fields as measured by wiring configuration and leukemia risk — is buried in the middle of the list:

> We conclude that our data offer no support for a relationship between measured electric field exposure and leukemia risk, little support for the relationship between measured magnetic field exposure and leukemia risk, some support for a relationship between wiring configuration and leukemia risk, and considerable support for a relationship between children's electrical appliance use and leukemia risk.

The information about measured electric field exposure and leukemia is misleading, since nobody has ever implicated electric fields as causing cancer; only magnetic fields are a cause of cancer. (The term *electromagnetic fields* refers to electric fields and magnetic fields taken together.) The unsuspecting reader might easily be tricked into thinking that electromagnetic fields (that is, both the electric fields and the magnetic fields) had been given a clean bill of health in the report, which is not the case.

This misleading statement is repeated on the following page: "The USC team found no association between childhood leukemia and measures of electric field exposure." Obviously, this material is meant to be confusing.

The report also confuses by understatement: "There have also been observations in several studies of an increase in cancer in workers in electrical occupations." In fact, by the time this report was written, over thirty occupational studies had found a significant increase in cancer in electrical workers, with many of them reporting as much as twice the number of cancer deaths in workers with high EMF exposures than in any other occupation.

The EMF cover-up continued into the nineties. The utilities took a page from Big Tobacco's playbook using the tactics of disseminating misinformation, belittling the EMF health effects research with cries of "junk science," and calling for a cut in EMF research spending.

In the late nineties Science by Public Relations took hold of the EMF issue. Just when most people were beginning to understand that EMFs had genuine health risks, two major reviews were published that assured the public, wrongly, that EMFs were safe. The vested interests had pulled a fast one on the media. All it took was a skillfully worded press release.

THE NAS-NRC REVIEW

In November of 1996, the National Research Council of the National Academy of Sciences (NAS-NRC) reported on a three-year meta-analysis of eleven EMF epidemiological or epi-studies. (A meta-analysis combines data from a number of studies and examines them in new ways.) The widely reported NAS findings stand in direct contrast to the long-standing EPA EMF report, which has still never been released. Authors of the EPA report criticized the NAS for misleading the public.

A headline in Newsweek (Nov. 11, 1996) is representative of coverage of the NAS report. "ELECTROMAGNETIC FIELDS BEAT THE RAP." The full page article goes on to trumpet the fact that the prestigious NAS-NRC found "the current body of evidence does not show that (EMF) exposures present a human health hazard." The mood of the Newsweek coverage was reflected overall in the nation's press, which reported unanimously that the NAS had found no evidence of an EMF-cancer link.

One publication did mention that the NAS had detected some weak connections between EMFs and childhood leukemia, in the form of high-current configuration (HCC) Wire Code houses in the residential studies and went on to mention that "doubters did remain." What no one mentioned was that the doubters were several well-known scientists on the NRC panel, who stated publicly that the committee did not, in fact, find that EMFs are safe — and that the committee did not write the NAS-NRC press release.

Three scientists in particular, researchers with impressive EMF backgrounds — Drs. Luben, Stuchly, and Anderson — went public with their criticism of the report. "We don't think the EMF issue is settled," Dr. Luben stated unequivocally, pointing out that some important studies were expected to report in the near future, studies that, if included, might change the analysis of the NRC review.

In a press release dated October 17 and published in the November, 1996, Bioelectromagnetics Society (BEMS) newsletter, the three scientists pointed out that the NAS-NRC found a link between EFMs and childhood leukemia and recommended further studies. The NRC had found an EMF-cancer link when EMF exposures were rated by the Wertheimer HCC Wire Codes surrogate — but the committee chose to dismiss it. Their reason, apparently, was that there was no clearly established link between measured fields or a dose/response finding, and the lack of CONSISTENT AND CONCLUSIVE proof. The committee also ignored the myriad positive occupational studies linking EMFs and cancer.

Dr. Luben wrote: "This report documents that EMF exposure produces a number of biological effects, both on cells in the lab and on animals, that could play a role in cancer development" and warned against taking this to mean EMFs are safe. He pointed out that it took fifty years to prove cigarettes caused cancer, even though there was plenty of epidemiological and statistical evidence of the relationship.

The scientists called for more research — as had the NAS review. Far from calling for an end to EMF research — as most of the media coverage, driven by the press release, implied — the NAS report had made a number of specific recommendations for further EMF studies.

THE NCI STUDY

On July 3, 1997, the National Cancer Institute reported on a study of EMF/Leukemia (Linet et al.). The NCI press release, headlined "Study Finds Magnetic Fields Do Not Raise Children's Leukemia Risk," provides a fascinating study in the art of spin control.

Although the study itself clearly states in four places that increased cancer risks are associated with EMF exposure, the press release begins with: "A comprehensive study by researchers from the National Cancer Institute (NCI) and the Children's Cancer Group (CCG) found no evidence that magnetic fields (EMFs) in the home increase the risk for the most common form of childhood cancer."

The press release continues, "Researchers found that, in general, children who lived in homes with high measured magnetic fields were not significantly more likely to be diagnosed with acute lymphoblastic leukemia (ALL) than children living in homes with low MFs. Nor was ALL found to be more likely among those whose homes were classified in high categories of 'wire code' . . ." The press release was a smokescreen, but one that worked. News that the NCI had reported that EMFs were safe was widely reported by the national press, prompting papers like *The New York Times* to report "Big Study Sees No Evidence Power Lines Cause Leukemia." The news coverage of the study's release came straight from the press release, not from the study itself. Even the *New England Journal of Medicine*, where the NCI study was published, reflected the PR version of the research. Accompanying the Linet report was a letter by the associate editor of the Journal, Dr. E. Campion, calling for an end to EMF research.

Dr. Raymond Neutra, commissioner of the California EMF program, wrote to the state Public Utilities Commission and the Journal to object to the editorial. Neutra pointed out that a number of large, important EMF childhood leukemia studies — a Swedish study, two German studies, two Canadian studies, and a British study — are in the pipeline and will be reporting in the near future. So the EMF issue has hardly been put to rest. His own California Health Department EMF program will be conducting a meta-study in 1999.

"I disagree with Dr. Campion's assessment that the study by Linet et al. was so superior in QUALITY and SIZE that it has, on its own, laid to rest the hypothesis that EMFs cause childhood ALL. Linet et al. . . . show results for 408 cases and controls."

Dr. Kjell Hansson Mild, Dr. Barry Wilson, and Cindy Sage (an EMF consultant) issued a press release that said, in part, "On the basis of this new study, some scientists and media . . . have repeated the questionable claim that the link between EMF exposure and cancer risk is no longer an issue and that further research is unnecessary. Such statements, based on a single study, are disturbing. More disturbing still is the fact that the data presented in the Linet study do not support the assertion that no link exists. Even a cursory review of the main data set shows a 53% increase above 3 mG and a more than 600% increase at exposures between 4 and 5 mG."

Ross Adey pointed out that, if they had included the 3-mG level in their cutoff, the conclusions would have been EXACTLY THE OPPOSITE — that there IS a significant risk. David Savitz used 3 mG in some of his work. Selection of 2 mG is "completely arbitrary," according to Dr. Adey.

In fact, the study data found that the cancer patients as a whole had higher EMF exposures: Compared to children living in homes with a field of less than 0.65 mG, children exposed to fields of 2 mG or more had a cancer rate twenty-four percent higher, while those living in homes with a field of 3 mG or more had a seventy-two percent increase in risk. (The second finding was statistically significant.) The NCI study does indicate that children with magnetic field exposures below 2 mG do not have an increased risk of developing leukemia.

But why did Dr. Linet's team choose the cutoff point of 2 mG — the point at which the effects disappeared — as the cutoff point in their study? It is impossible not to question this decision. Why mislead the public by ignoring the point at which there were clear risks? Why did the National Cancer Institute fail to warn the public about the hazard?

One answer is competition for a limited amount of research money. But, another is the existence of NCI hidden agendas that have been known to influence policy decisions. The NCI's entire history has been marked by the agency's unwillingness to make full public disclosure on any number of public health issues.

NCRP

The public has a difficult time sorting out the complexities and politics of the EMF issue, so it's no surprise that efforts to mislead it are often successful. The handling of these two big studies did its damage; right now, many people really believe they don't have to worry about EMFs — when, in fact, a number of compelling large studies have been reporting that EMFs are hazardous to your health. The positive studies coming out in the late nineties were among the most compelling ones, yet many received little press coverage.

First among these to warn that EMF exposures are hazardous to your health came from the prestigious National Council on Radiation Protection (NCRP).

NCRP's Scientific Committee 89-3, a panel of eminent scientists, had been charged by Congress to conduct a seven-year review to determine whether there was a significant EMF risk and, if so, to develop some good national EMF guidelines to protect the public. Late in the summer of 1995, a draft of the executive summary containing NCRP's recommendations on EMF exposure was leaked to the press.

In contrast to the NAS and NCI, the NCRP recommended a national public exposure limit of 2 mG because higher levels of exposure increase the risk of cancer. The review cited "epidemiological findings that suggest significantly enhanced cancer risks, particularly for childhood leukemia, in ambient power frequency fields exceeding 2 mG."

The report goes on to state: "In key areas of bioelectromagnetic research, findings are sufficiently consistent and form a sufficiently coherent picture to suggest plausible connections between ELF EMF exposures and disruption of normal biological processes . . ."

NCRP also recommended that new day care centers, schools, and playgrounds not be built where ambient 60-Hz magnetic fields exceed 2 mG; that new housing not be built under existing high-voltage transmission lines; and that new transmission lines not be built in locations where they'd produce fields of 2 mG and higher in existing housing.

Just as the 1990 leak of the EPA EMF review had caused a flurry of spin control from the industry, the same was true after the leak of the NCRP report. Spokespeople from utilities claimed this was only a draft that didn't have the backing of the committee.

This was denied by Dr. Richard Luben — a longtime researcher in electromagnetics who is on SF 89-3 and helped write the executive summary. According to Luben, the executive summary had agreement from the scientists on the panel and, furthermore, after the committee's review, he believes that living close to power lines "is likely to be associated with increases in childhood leukemia." At hearings in Australia about a major power-line project, Dr. Luben testified that, "The correlation between power lines and leukemia is statistically supportable. There are possible mechanisms based on both animal and laboratory results that suggest cancer-causing or cancer-promoting activities of EMFs," adding, "This leads me to feel there is some reason for caution." Luben also said that the NCRP committee felt that the (1990) EPA report's conclusion that EMF exposure was a possible potential carcinogen "was a valid conclusion."

OTHER MAJOR STUDIES

In 1996 and 1997, a slew of major studies — including some huge 'meta-studies' (analyses which combine data from a number of past studies and examine them in new ways) — reported positive findings on the link between EMF exposure and cancer rates. Although none of these got much coverage in the press, they punched some important new data into the EMF record.

A brief review of these studies includes:

- ALBOM, FEYCHTING & OLSEN meta-analysis that combined the data from two large Scandinavian childhood epi-studies and confirmed the childhood cancer risk from EMFs. The study reported increased childhood cancer with exposure to EMFs, with a dose-response relationship: the higher and longer the exposure, the greater the risk beginning at 2 mG where the risk was doubled. (November, 1995).
- KHEIFETS & BUFFLER conducted an EPRI-sponsored meta-analysis that reported increased cancer risks for electrical workers. The study reported a twenty percent increase in brain cancer with high EMFs. (Most of the studies used in this meta-analysis had reported increased brain cancer risks.) (December, 1995).
- MILLER did a meta-analysis using the Ontario Hydro worker data and found increased risk of leukemia — up to eleven times the expected risks — linked to ELECTRIC FIELD EXPOSURES. (January, 1996).
- Another large study done that same year by GUENEL, who revisited the Electricite de France utility worker data, also found increases in brain

cancer associated with electric field exposure. (These two studies have revived interest in revisiting the issue of electric field vs. magnetic field health effects; critics are skeptical, for one thing, because of the ease with which electric fields are shielded.)
- In '1996 FEAR looked at 372,000 cases from the cancer registry in England and found greatly increased brain cancer and leukemia risks for electrical workers.
- CHUNG YI-LI, GILLES THERIAULT, RUEY LIN did a large residential study in Taiwan and reported increased leukemia risks in adults with EMF exposures. The study found a dose-response relationship: for exposures of 2 mG and higher, a forty percent greater leukemia risk and a seventy percent greater risk for ALL. (May, 1997)
- FEYCHTING, FLODERUS & FORSSEN (in Sweden) studied 400,000 adults' MF exposures at work and at home and reported that subjects with high combined exposures had four to five times the leukemia risk. (July, 1997)
- That same summer, SAVITZ & LOOMIS conducted a meta-study of 140,000 electrical workers and found an association between EMF exposure and ALF (Lou Gehrig's Disease). This analysis re-examined the EPRI utility data from their earlier study in 1995 that had reported a significant link between EMFs and cancer.
- In May, 1997, there was a pivotal study that did catch the attention of the world's press: in Australia, REPACHOLI reported double the lymphoma in mice that were exposed to microwaves that mimicked a digital cell phone signal. (See Chapter 8 for more on this.)
- In 1996, STUCHLY & BYUS had confirmed their earlier findings that EMFs were a co-promoter (with a known tumor promoting chemical) of cancer in mice.
- (For more studies, see APPENDIX A.)

A great deal of important EMF research is still forthcoming, including these large reviews and meta-analyses:

- The DOE/NIEHS (National Institute of Environmental Health Sciences) $65 million RAPID (Research & Public Information Dissemination) EMF study, mandated by Congress in 1992. RAPID is co-funded by the government and industry. RAPID-funded research includes studies on nearly all the major areas of EMF inquiry: animal and in vitro EMF mechanism studies, a breast cancer study, research on risk assessment, and an epidemiology meta-analysis that will include the NCI data. Due to report at the end of 1997, RAPID has been buffeted by politics, funding

problems (nearly half the industry money hasn't come through), infighting, and delays. The future role of the federal government and its funding of EMF research is riding on what RAPID reports. Right now it looks as if the government is shutting down all its EMF research. The EPA has even disconnected its EMF hot line.
- The worldwide EMF Health Effects Project sponsored by the World Health Organization (WHO).
- The California Department of Health Services has an EMF program, headed by Dr. Ray Neutra. The $7 million project, begun in 1994, is easily on a level with federal EMF research. The goal of the project is to shape a reasonable approach to potential EMF hazards. The scope of the project is enormous: laboratory research, an epi-study meta-analysis, exposure assessment, mitigation research, and education. Part of the project focuses on measuring EMFs in schools, determining their sources and mitigating them — with an eye toward making policy decisions informed by mitigation costs. (For more on this project see Chapter 5.)

People are also waiting for results from a number of key breast cancer studies, among them the Seattle NCI study by Drs. Scott Davis and Richard Stevens (More on this in Chapter 4), a study by Dr. Stephanie London at the University of Southern California, a large Swedish study by Feychting et al., and replications of Wolfgang Loscher's work. (According to Dr. Stevens, a pioneer in this work, "The idea (that MF exposure is linked to breast cancer) is really out there now and is taken seriously."

The NAS and the NCI may have told the public EMFs are okay, but reports of the demise of the EMF health issue have been widely exaggerated. THE FINAL RETURNS ARE NOT IN YET ON THE HAZARDS OF EMFs, not by a long shot. Despite the fact that vested interests would have you believe otherwise and that many scientists, for reasons of their own, are silent about what they know, while others cite uncertainty as reason to abandon search — the evidence that EMFs are hazardous to your health is increasing all the time, and the issue won't go away.

Chapter 4

What the Studies Tell Us

> *I have seen the change from belief there was no effect — couldn't possibly be — to an acknowledgment that there is an effect on biological systems. And some possible effects on embryogenesis and fetal maturation, as well. I've always said, exposure should be limited and people should keep aware of what is coming out. You should have a judicious attitude. For heaven's sake, don't stand under a bloody power line.*
>
> DR. WENDELL WINTERS,
> University of Texas Health
> Sciences Department

If you wonder whether EMFs are hazardous to your health, turn to the back of the book and read APPENDIX A. This appendix is a compilation of decades of positive studies — studies reporting significant bioeffects when people, animals, or cells were exposed to EMFs at 60 Hz or RF/MW levels. You'll see that for the past two decades, a growing body of scientific evidence has linked EMFs to increased health risks. This chapter will help you interpret those studies within the context of the overall health issues concerning EMFs. Then you can decide for yourself.

According to a large, respectable sector of the scientific community, which includes researchers and physicians who have been working in the field as well as public health officials familiar with the data, enough studies have been done that show EMFs are dangerous. Although it's certainly true that we need to learn more about how EMF exposure affects us, we already know a good deal. Dr. David Carpenter, dean of the State University of New York School of Public Health and chief of the New York State Power Lines Project, put this whole matter into perspective:

On the basis of the scientific studies that have been well done to date, I view the state of the evidence to be very clear on exposure to electromagnetic fields and cancer. It's very difficult in this business to define the point in time when the evidence goes from being strongly suggestive to "proof." Take cigarettes and cancer as a case in point. Cigarette smoking is a good example of the kinds of problems we run into here. When was proof obtained that smoking caused cancer? It's hard to pinpoint. The debate raged on for years and years. In fact, in cases where people are connected with the cigarette industry, it's still going on. Right now, I'd say it's appropriate for individuals to take whatever steps they might without too much expense or disruption in their lives to reduce their overall exposure to EMFs. We certainly don't need additional studies to tell us what the dangers are.

The basic evidence that has established the relationship between electromagnetic fields and cancer consists of:
1. Residential epidemiological studies that examined the incidence of cancer cases in communities reported that people exposed to power-line EMFs over 2 mG have a greater risk of cancer. Among these, almost a dozen childhood cancer studies have reported increased risks of leukemia, lymphoma, and brain cancer with EMF exposures of 2 mG and higher. Some studies found a dose-response link, i.e., the higher the exposure and the longer its duration, the greater the risk.

In many of the epi-studies, field calculations based on wire code configurations and estimates of occupational exposures were linked to significant cancer increases — but measured fields were not. This suggests that measurement of EMFs is not yet accurate, not, as vested interests would have it, that there are no risks. The question is, What is being measured? Ambient EMFs have a number of variable components. It may be that today's time-averaged or one-time measurements don't catch the particular aspect of the magnetic field (like transcients or high peaks) that may turn out to cause the increased risks. The point is, just because the science doesn't yet know how to capture it, doesn't mean there isn't an association.

2. Occupational studies that have consistently supported the residential studies and given them weight, by reporting that workers with high EMF exposures have increased risks of cancer, in particular leukemia and brain cancer.

3. In vivo (live animal) studies that have found EMF exposure increases tumor production and other carcinogenic effects, bearing out the human studies.

4. In vitro laboratory experiments that have shown that EMF exposures have effects on cells and tissues. This work supports the epi-studies, reporting tumor formation and other mutagenic effects in cells and tissues exposed to EMFs. EMF laboratory studies have produced findings consistent with an initiation-promotion model of cancer, suggesting that EMF exposures are involved in the production of cancer.

Early studies that examined only electric fields were used as propaganda, as Dr. Robert B. Goldberg pointed out in an article in Medical Hypotheses:

> Most of the large-scale animal exposure experiments conducted before the late 1980s were designed to assess hazards associated with high-voltage power lines and utilized electric field exposure systems with little or no magnetic field component. Consequently, the generally negative results of pathology screening are of questionable applicability to concerns raised by the epidemiological data. (Nonetheless, special interests will continue to refer to these very studies to counteract the positive effects of newer studies that have focused on the magnetic field exposure.)

The majority of these newer studies linked MAGNETIC FIELD EXPOSURES to increases in cancer. Recently, however, two studies have linked increases in cancer to exposure to ELECTRIC FIELDS. (This electric field finding has to be examined carefully; many people are skeptical about it because, for one thing, electric fields are so easily shielded.)

To understand what happens to living things when they're exposed to a magnetic field, particularly when that field causes cancer, we need to know a little more about carcinogenesis — the production of cancer.

SUPPORTING EVIDENCE OF CARCINOGENESIS

In 1989, the EPA issued a 150-page document that is full of evidence of the potentially carcinogenic effects of EMF exposures. Some of the language in the report seems unnecessarily equivocal: "In view of the laboratory studies, there is reason to believe that the findings of carcinogenicity in humans are, at least, biologically plausible."

Bear in mind that a great many experts today believe that ambient magnetic field exposures result in increased cancer. Besides the growing epidemiological evidence, a large number of laboratory studies have reported that the proliferation of cultured cancer cells greatly increased when the cells were exposed to power-line-strength magnetic fields. In some studies cells doubled or tripled in a matter of a few hours. Studies have also found that fields inhibited the cytotoxicity of the natural killer lymphocytes and prevented them from destroying tumors. (The fact that EMFs can somehow prevent our bodies' defenses from stopping cancer once a cell mutates is a serious enough allegation to make us want to avoid them.)

Most of these investigators believe magnetic fields act as a cancer promoter, not an initiator. This is partly due to the generally held belief that the extremely low-frequency (ELF) radiation of a 60-Hz electromagnetic field doesn't have enough energy to produce mutagenic effects. That is, it can't actually enter a cell and break the chemical bonds in the DNA (deoxyribonucleic acid), causing mutations, as a cancer initiator must.

Once we start to think of EMFs as cancer promoters, a number of things become clear. For example, if we understand magnetic fields as promoters, we can put to rest the whole argument critics use about chemical confounders — that other chemicals could be responsible for the reported effect — in the EMF occupational studies. For this model, the fields and other chemicals would work together to cause the disease.

A paper by Indira Nair and M. Granger Morgan, in the August, 1990, IEEE (Institute of Electrical and Electronics Engineers) Spectrum, stated: "Even if the cancer source in electrical occupations is something else, such as a chemical, *the fields may enhance its action*" (emphasis added). Of course, when you understand the workings of the disease, this is every bit as worrisome. It means that once cancer is started by something else, exposure to EMFs can make it progress more rapidly.

Many of the reported EMF effects point to a direct link between the fields and the biochemical process of the disease, at the cell level. These effects involve changes in mitosis or cell proliferation, hormone production, calcium efflux, ODC (ornithine decarboxylase) activity, and melatonin production.

CANCER TERMINOLOGY

Cancer is a disease marked by the production of malignant tumors with virtually unlimited growth. *Carcinogenesis* is the process involved in the production of cancer. Certain kinds of *ionizing radiation* — gamma rays, nuclear radiation, and x-rays, all of which can get inside cells and break chemical bonds — and particular chemicals called *carcinogens* — like cigarette smoke and asbestos — are known to cause cancer. Thus, these chemicals should be avoided to reduce the risk of the disease.

The process of malignant tumor production in cancer is actually just the process of normal cell growth gone seriously awry. The normal process of cell growth and differentiation — *mitosis* — occurs in all living things as they reproduce or produce new cells to replace dead or worn-out cells in their tissues and organs. During mitosis, cells divide and produce two new identical cells or *clones*, called *daughter cells*. Prior to mitosis, the parent cell must duplicate its DNA molecules. (DNA is located in the cell nucleus and contains all the genetic information for the cell.) Each daughter cell then receives a complete set of identical chromosomes from the DNA in the nucleus of the parent cell. These chromosomes contain the *genes*, which are a repository for the genetic code or a blueprint for the inherited traits and characteristics of an individual. Genes control specific cell traits, like hair color, or direct the functions of particular cells and organs.

In this way, all the inherited traits and cell instructions are passed on during mitosis by RNA (ribonucleic acid) molecules that have been produced by the DNA and contain all its chemical instructions. These RNA *messenger molecules* carry that genetic message across the cell membranes and into the nucleus of the daughter cells during a process called *transcription,* or *gene expression.*

This same process of cell division is involved in cancer or tumor formation. A tumor is simply a cell whose reproduction has somehow gone out of control. Tumor cells clone wildly and spread, or *metastasize*, throughout the body as the disease progresses. Although a lot about the disease is still a mystery, we do know that this process of uncontrolled cell division is triggered by contact with either a chemical agent or ionizing radiation. The progression of cancer is understood to be a two-stage process involving initiation and promotion.

THE TWO STAGES OF CANCER PRODUCTION

STAGE ONE: INITIATION

The first stage of cancer occurs when the DNA in the cell nucleus is somehow damaged by nuclear radiation or X-rays or a chemical carcinogen like dioxin. This exposure causes the cell to *mutate* — that is, to change. At the point of mutation, the DNA genetic data in the cell have been irreparably damaged. The cell's ability to control its cell division may also be damaged. An agent that damages the DNA in a cell, causing it to mutate, is called an *initiator*.

Once such a mutation has occurred, the cell has become a cancer cell and is capable of passing on the mutation. However, that will not necessarily occur. Under normal, healthy circumstances, the damage can be repaired or controlled by the body's immune system. During Stage One, the mutated cell is said to be precancerous, or *neoplastic*.

The precancerous cell will have to undergo a second separate exposure — this time to a carcinogen called a *promoter* — before it will develop into a tumor. If the mutated cell never comes into contact with such an agent, the person will be in no danger of getting the actual disease.

STAGE TWO: PROMOTION

This stage of cancer begins when the precancerous cell is exposed to a second carcinogen, a promoter. A *promoter* is an agent that causes a cancer cell to develop into a tumor. (An agent — for instance, a particular chemical — is either an initiator or a promoter, not both; by definition, a promoter can't initiate cancer.) It is the exposure to the promoter that causes the mutated cell to undergo unchecked cell division and form a malignant tumor. When this happens, we say that a person has cancer.

What exactly causes this uncontrolled cell reproduction isn't understood, but the most widely accepted hypothesis is that something occurs that either changes the DNA blueprint on the transmitter RNA or changes the normal transmission of messages from the cancer cell to its daughter cells. Promoters are also known to attack the specialized lymphocytes in the immune system that would normally kill tumors — the cytotoxic T-lymphocytes, or Natural Killer (NK) cells — and prevent them from doing their job. (Inhibiting the cytotoxicity of the lymphocytes.) In any case, it's important to understand that without exposure to a promoter, cancer will not proceed.

Quite a bit of time may pass between these two stages in cancer production. This time is know as the *latency period*.

CHANGES IN CELL PROLIFERATION

A handful of compelling new studies have found that EMFs and certain carcinogens are CO-PROMOTERS of tumor growth. This research shows significantly increased tumor growth in animals that were exposed to EMFs and treated with tumor promoting chemicals.

Prior to this, the greatest source of information on cellular effects of EMF comes from in vitro research on cell proliferation. Various combinations of magnetic field strengths and frequencies have resulted in a number of important changes observed at the *cell membrane*. The cell membrane is the site where messages from the brain or the DNA are passed on to the nucleus of the cell to be translated into cellular activity. For instance, during the process of transcription, RNA molecules carry messages from the DNA across the membrane to the cell nucleus, which results in cell division. Cancer promoters are known to cause changes in signals or messages that are being carried across the cell membrane, and one obvious effect of such changes could be increased cell proliferation.

What The Studies Tell Us

A few studies on plant and animal cells have reported changes in the chromosomes — that is, *chromosomal aberrations (CAs)*, or *chromosomal breaks*. Some of these CAs have occurred when cells were exposed to a combination of ELF fields and microwaves. There have also been studies that found chromosomal aberrations in cells of workers who had been exposed to high levels of EMFs in their work, in particular switchyard workers.

On the other hand, any number of studies have reported changes in rates of DNA and RNA synthesis, RNA transcription, and enhanced gene expression as a result of EMF exposures. Such changes in gene expression are likely to result in changes in cellular activity.

Dr. Reba Goodman, of Columbia University, and Dr. Ann Henderson, of Hunter College, have conducted a large body of work with EMF effects on cells. Some of their studies suggest, but don't prove, possible MF effects on the genes themselves. Dr. Goodman explained, "Based on what we do, we could not say cancer can be induced by exposure to EM fields. We view the fields as a stressor. But there is definitely something here that's turning genes on and off out of sequence. Exposed genes become activated and controls do not." (Stress, in the sense that it's being used here, is known to cause changes in tumor growth in laboratory animals.)

Reviewing in vitro studies for a 1991 U.S Department of Health and Human Services workshop on the "Health Effects of Electric and Magnetic Fields on Workers," Stephen F. Cleary, Ph.D., wrote:

> In general these data indicate that the 60-Hz electric field effect on proliferation depended upon the mitotic status of the cell population during exposure. This finding is potentially significant since it suggests specific interaction of LF EM fields with the mammalian cell cycle. The dependency of the proliferation effect of EM fields on factors such as (this), if not taken into account in the design of in vitro studies, could result in highly variable or contradictory results.

The fact that the variation of effects seems to be tied to the mitotic status of the cells could shed some light on why children are so susceptible to EMFs: Their cells are already undergoing rapid growth, compared to adults' cells.

Dr. Cleary goes on to explain that the EMF effect seems to depend on some other factors as well: "It may be concluded that low-intensive LF EM fields modulate proliferation of normal as well as transformed mammalian connective tissue cells in vitro. . . . The magnitude of the proliferative response was dependent upon EM field intensity, exposure duration, and cellular and extracellular factors."

CHANGES IN HORMONES

Hormones play a key role in regulating most of the functions of our cells, tissues, and organs. This regulatory activity occurs at the cell membrane. Hormones control critical bodily processes like cell growth and differentiation, reproduction, and the immune response. In general, they work together and are subject to the maintenance of a delicate biochemical balance. EMF exposures have been found to alter the production of a class of critical hormones called *neurotransmitters* that carry nerve impulses from the brain across the cell membranes. This could have important implications in cancer, as well as other diseases.

To transmit instructions from the central nervous system, neurotransmitters bind or interact with *receptors* on the cell membrane (structures on the membrane that transmit messages from outside to the nucleus of the cell). The number of receptors on a cell membrane can vary; one way scientists determine how much activity is going on inside a cell is by counting the receptors. Chemicals that promote cancer are known to increase the number of receptors on a cell. Studies have shown EMF exposures greatly increase the number of receptors on cells.

CHANGES IN CALCIUM EFFLUX

Receptors following the instructions of the hormones release *calcium ions* that are stored in the cell membrane. These calcium ions are the primary messengers of our cells; their job is to carry signals back and forth across the membrane. In this way, calcium exerts a critical regulatory activity on normal cell growth and differentiation and tumor production as well. Many studies of cells exposed to magnetic fields have reported changes in this calcium flow, or *flux*. In some studies combinations of AC and DC fields have caused the effect. In vivo studies of live animals have also reported increased calcium flux in the animals' brain cells after they were exposed to EMFs. All these findings suggest changes in cell reproduction such as those associated with unchecked tumor growth.

Calcium flux, in turn, stimulates production of a group of enzymes called *protein kinases*, which have a key role in the regulation of many cellular functions, including cell proliferation. Studies have reported decreased protein kinase activity in human lymphocyte cells exposed to EMFs, which suggests that the fields suppress the immune system. Since kinases stimulate cell proliferation, their absence in the lymphocytes could, in effect, halt production of the tumor-killing cells when the body is being attacked by cancer.

INCREASED ODC ACTIVITY

The metabolic enzyme ornithine decarboxylase (ODC) is thought to be a key player in tumor co-production because of the fact that it controls the rate of cell growth and proliferation. Cells can only reproduce as fast as ODC is synthesized. ODC levels are a reliable marker for malignancy because, just as the enzyme is always present during normal cell reproduction, it's present in far greater amounts during tumor production. ODC is also known to increase in the presence of chemical cancer promoters. Many studies have linked increased levels of ODC to EMF exposures. One researcher (Byus, University of California at Riverside) who found EMF exposure greatly altered ODC activity in cells, suggested that 60-Hz EMFs function as tumor promoters.

INHIBITION OF MELATONIN

The finding that MF exposure inhibits the production of the hormone melatonin may be one of the most crucial results of bioeffects research — especially since it seems to point to an EMF mechanism that could cause cancer.

Melatonin, a hormone produced by the pineal gland, regulates *circadian rhythms*, or what is commonly known as our *biological clock*. Circadian rhythms control our sleep-awake cycle, moods, and task performance. Reduced melatonin levels produce depression, severe mood changes, and certain psychiatric disorders. Pineal production of melatonin normally occurs at night; it's controlled by the amount of light detected by the retina.

Melatonin is also a known inhibitor of tumor growth; it increases the cytotoxicity of the natural killer lymphocytes. Doctors report decreased melatonin levels in their cancer patients, and melatonin is used in chemotherapy to fight the disease — particularly breast cancer and prostate cancer. (It's interesting to note that blind women have greatly increased melatonin levels and very little breast cancer.)

Experiments have shown that magnetic field exposures lower melatonin production and interfere with its ability to inhibit tumor growth. Studies have linked melatonin reduction to increased mammary cancer in animals, increased proliferation of cancer cells, and impaired immune functions.

Dr. Russell Reiter, author of *Melatonin: Your Body's Natural Wonder Drug*, explains, "If you get a depression of melatonin from any of these fields, you increase the damage to various organs — particularly the brain. It's important that a lot of the epidemiology concerns brain cancers.

"There's no doubt that certain electric fields and magnetic fields influence the production of melatonin. And these fields are within the range of frequency strengths to which we're generally exposed.

"I would predict other subtle central nervous system effects, as well. Think about how you feel when you experience jet lag. The effect of all this is not unlike jet lag — you feel a general lethargy. Your working efficiency would be down. A common effect from reduced melatonin is an inability to sleep, which adds to the general feeling of exhaustion.

"Then think of a major company that employs 1,000 workers. If they all arrive at work each day not at their peak because EMFs are interfering with their melatonin production, they're not going to perform well. They're going to require more breaks. Be out sick more, and maybe have more accidents. What I'm saying is, this could cost a large company millions. And depressed melatonin levels are associated with depression, what is known in psychiatric circles as 'low melatonin syndrome'."

Dr. Richard Stevens of Battelle Pacific Northwest Labs believes increased EMF exposures account for the rise in cancer that has been linked to industrialization. He points out that electrification goes hand in hand with industrialization and cites occupational studies reporting increased breast cancers in workers with high EMF exposures. As early as 1987, Dr. Stevens wrote, "The use of electric power may increase risk of breast cancer. The hypothesis is based on experimental evidence that shows an effect of light and ELF electric or magnetic fields on pineal melatonin production and on the relationship of melatonin to mammary carcinogenesis." In a later paper, he said, "Suppression of pineal function has been implicated in the etiology of several types of cancer, including breast, prostate, ovarian, and melanoma . . . EMFs can suppress pineal function." At this writing Dr. Stevens is wrapping up a major NCI-funded study of breast cancer and electricity. There are also a number of similar studies going on around the country.

Other potential health risks from EMFs may be related to melatonin. Dr. Robert Becker believes that central nervous system effects from lowered melatonin are going to be very important and that the large increase in students with learning disabilities and attention deficits in the U.S. may be linked to EMFs.

REPRODUCTIVE HAZARDS

There have also been studies showing a clear relationship between magnetic field exposure and reproductive hazards. Studies have reported increases in miscarriages and birth defects as a result of exposure to fields. In this case, as well as in the case of cancer, the goal of protecting the public has been ignored in a concerted effort to simply make the findings go away. Although some of these studies have been reporting since the early sixties, they've been routinely covered up by the military and major industries that have a vested interest in unhampered technological expansion.

THE SCIENTIFIC METHOD

The scientific method is built on the concept of scientific consensus — the idea that a number of independent researchers are going to be able to come to some sort of agreement about a question of scientific interest, such as whether EMF exposure can cause biological effects. This notion of obtaining clear consensus has put many EMF issues on hold as controversy about the subject continues. Scientists as a group are notorious for disagreeing, and one of the things they're likely to dispute is what a certain scientific study really means, which is what has happened with EMF research.

The scientific method relies on three systematic procedures: (1) the formulation of a hypothesis to be tested; (2) the carrying out of experiments and observations to collect data relating to this hypothesis; and (3) replication of these tests to see whether the results are consistent. (Step 3 may be carried out by the original researcher or by others.) Next comes the peer review: When a study is completed, it must undergo the formal scrutiny of other scientists, who either criticize or accept it.

The point is, this whole procedure takes time — often years and years. And there's no predetermined deadline for the scientific review period to come to an end. All that is needed to prolong the process is the call for more research. As you can see, this process can present a real problem when consensus is needed on a subject of some urgency, like whether or not EMFs are a public health threat.

The very idea of EMF radiation causing dangerous bioeffects threatens two major scientific paradigms: one for biology, the other for physics. These basic paradigms (or tenets) hold that it's impossible for nonionizing radiation to affect biological systems. The scientific community is saying, "We don't know how this could happen — so it doesn't!" Furthermore, scientists don't like to give up their paradigms, even though the history of science is simply a record of replacing one paradigm with another.

WARNING: the electricity around you may be hazardous to your health

M. Granger Morgan, head of the Department of Engineering and Public Policy at Carnegie Mellon University, addressed this question when he told a conference of physicists, "In my mind, the many positive studies in the refereed literature affirmatively resolve any question of whether effects exist."

What can society do if people feel an urgent need to have some scientific answers in order to protect themselves from a possible health risk and the scientific verdict is not forthcoming? In the absence of clear scientific proof, or in the face of a continuing scientific controversy, society often acts by reviewing all the available data carefully, then making a judgment call informed by the weight of the evidence. Many experts like Dr. David Carpenter, of the New York State Department of Health, feel the time has come to take a weight-of-the-evidence approach to the question of the health risks of EMFs.

"On the basis of the scientific studies that have been well done to date, I view the weight of evidence to be very clear on exposure to electromagnetic fields and cancer," Dr. Carpenter explained. "It's very difficult to define the point in time when the evidence goes from being strongly suggestive to proof. Right now, I'd say it's appropriate for individuals to take steps to reduce their exposures to EMFs."

Dr. Samuel Milham, ex-chief of the Washington State Department of Public Health, agrees. "There have been enough studies that found links. You can't toss a coin 40 times, get heads 35 times, and call it coincidence." (When concerned citizens called Dr. Milham's office, they were counseled to keep their EMF exposure below 3 mG.)

Despite the already strong evidence, the debate rages on, fueled by people with conflicts of interest who want to keep the public confused about whether EMFs are harming them and are using scientific arguments to do it. One argument that's repeatedly raised by special interests, to fend off regulation, is that there still isn't enough definitive scientific proof. It is true there are a number of legitimate unresolved questions — questions about doses, about the mechanisms involved when EMF affects us biologically, where and when and how exposure can make people ill. But, it may surprise you to learn that science still has similar questions about every single known carcinogen. The state of the science is weak when it comes to definitive data about the mechanisms of disease. Any medical researcher will tell you we don't really understand the mechanisms of a single chronic disease — and that includes cancer.

The same people who keep lobbying for more and more studies have some other strategies to keep the public in check. One is to mislead people about so-called "negative" and "positive" studies. A positive study is a piece of research that proves the hypothesis it set out to test. A negative study fails to find such a correlation. These two kinds of studies aren't generally given equal weight by the scientific community. Many negative studies are flawed or weak and they should be repeated. (And remember what Dr. Goldberg said: Most of the early studies were never designed to look for magnetic field effects.)

Another method of obfuscation used by detractors involves the scientific practice of reexamining questionable data by restructuring the data into new groupings, then taking a fresh look at them. In so-called "meta-studies," data from a number of related studies may be combined, then sophisticated computer analysis is applied on the enlarged data bank. This has been done to good effect with some EMF studies — as when Savitz reviewed cancer risks for particular electrical professions by combining mortality data from sixteen states. But, this method can also be employed to change or water down the results in a study, by tinkering with the data until they say what the researchers want. Studies can also be rigged by designing them to come up with negative results, then using them to contradict positive studies.

Confounders can also confuse the results of a study. They are variables that may cause the results without being identified in the research. The question of confounders is often raised with regard to the occupational studies, with the concern that other chemical exposures, rather than EMFs, are responsible for the reported cancer increases. This ploy is popular when people try to discredit these studies. In fact, all the later well-credited studies tested for confounders and ruled them out. (Nevertheless, special-interest spokespeople still continue to use this argument, like the utility company public relations man in California who tells people Savitz found heavier traffic near the high current configuration (HCC) homes in his study, which may just as well be the cause of the cancer increases. In fact, Savitz has gone on record saying that no other confounding was noted and that traffic density was associated with both power-line configurations and cancer occurrence, but the association wasn't strong enough to confound the association between power lines and cancer.)

There's another subtle point to understand when it comes to disease. There's a big difference between a *confounder* and a *cofactor*. In many diseases, two or more factors act together to cause the condition. These different agents, known as cofactors, must both be present for the disease to occur. In fact, this is believed to be the case with cancer, where an initiator and a promoter acting in concert cause the disease. If magnetic fields are, as is currently believed, promoters of cancer, then the existence of other chemical carcinogens in the data really would not disprove the role of EMFs in increased cancer risks.

Something else apologists for the electrical industry often do is treat epidemiology as though it were a poor relation to real science, implying that because some of the strongest EMF findings are only epidemiological, they needn't be taken seriously. Epidemiology — the study of the incidence, distribution, and control of diseases in the population — is taken very seriously because in the relatively short time that it's been around, about thirty years, it has made major contributions to the public health. For example, most of what we know today about heart disease, about the link between smoking and cancer, and about the hazards of many environmental toxins, came from epidemiological data. Epidemiological studies often point the way to needed laboratory studies, as has happened in the EMF/cancer arena.

One clear advantage epidemiological studies have is that they're already focused on cases of actual disease in the population, so they can have immediate public health implications. Much of our current regulatory action has been the direct result of epidemiological studies, as in the case of asbestos, arsenic, ionizing radiation, and smoking. Most of these things were regulated on the basis of the epidemiology alone, according to Dr. Martin Halper, director of the U.S. EPA Office of Radiation Programs. Dr. Halper also has said that the epidemiological data is very strong in the case of EMFs.

"If we make the case on EMFs, it'll be made on the basis of the epidemiology. People argue that the risk ratios that have been reported aren't high enough to warrant regulation. But, once we figure out how to measure the exact component of the fields that causes the effect, we'll probably get higher risk."

This argument — that the effects or the reported risks are small — is often used to mislead the public, when, in fact, all the EMF epidemiological studies probably underreport risks. The reason for this is the following. In a typical epi-study, two groups are investigated and compared — the subjects with the condition (in this case, cancer) and the controls, who don't have it. The problem is, in the case of EMF exposure, there are no real controls because, unlike most hazards, everyone is exposed to ambient EMFs. This automatically skews the data downward; it's fair to assume all the findings are larger than reported.

WHAT'S WRONG WITH THE STUDIES?

There are, however, other problems with the EMF literature that derive from limitations in the science. A glaring problem with many epi-studies is measurement, or dosimetry. No one has come up with a widely accepted, consistently reliable method to assess EMF exposures. Exposure metrics vary from study to study, which sometimes adds a note of confusion to an admittedly inexact science, making it difficult to compare different studies.

What The Studies Tell Us

As you read the various studies, you'll be struck by the variety of measurement methods employed by different researchers. Some studies used one-time spot measurements, some did twenty-four-hour exposures — and these may have been done either in the children's bedroom or at the front door of the house. Many studies used surrogate measures to calculate the fields, like the Wertheimer/Leeper Wire Codes where fields were determined by the wiring configurations: the thickness of the wires, the location of transformers and service drops, and so on. Some studies rely on historical electric company records of the probable loads — which may or may not be accurate. The time of exposure being measured also varies: one study may measure fields from the year of diagnosis, another from two years after diagnosis, or another may track the child's entire lifetime in a variety of residences.

Remember, the use of the word "actually" referring to measurements is fraudulent, as in: when they actually measured the magnetic fields they found no correlation. This bogus language is often used by vested interests to mislead the public into believing a negative study — when in fact the problem may have been that the method of measurement wasn't good. and it is misleading to imply that measurements done with dosimeters are automatically superior to surrogate methods of determining fields. Studies that measured fields (notice I didn't say "actually measured fields") are no more intrinsically reliable than studies that used surrogates.

The truth is, no study to date has used "actual" measurements, which would be only one thing — long-term measurements made at the time the disease was developing.

Research using surrogate measures has produced the most compelling data linking EMF exposure to cancer. Because of the uncertainties surrounding the whole question of EMF measurement, it is a mistake to discount this data out of hand.

Another problem concerns cut-off point. Arbitrary cut-off points can easily skew the results of a study, as we have seen in discussion of the NCI study in the preceding chapter. The findings would have been dramatically different if the cut-off point had been 3 mG instead of 2 mG! When you're designing a study, you can do many things to guarantee negative results. In EMF research, one of the things you can do is make the cut-off point so low and the control exposure so high that you get no effect. The study is designed to find nothing — and it succeeds.

This brings us to the issue of overall design of a study: what a researcher sets out to find — and what the study isn't supposed to investigate. Take, for example, the decision to measure electric field exposures, magnetic field exposures — or to combine the two. Another decision that can lead to significantly different findings which can then be used to mislead the public.

And, by the way, what about so-called Junk Science? Like Junk Food, we know it's bad for us — but what is it really? And how can we tell Junk Science from the real thing? So far as electromagnetic field research, Junk Science is a charge that's been leveled at the science by interested parties — the vested interests, the establishment, and ranks of lobbyists who invented the label to further obfuscate EMF concerns. It's a catchword that's being used to label any study these people want to vilify. Unfortunately, the label works; it has certainly been effective in getting EMF lawsuits thrown out of court.

THE PUBLIC HEALTH CONSIDERATION

The public health consideration about a particularly urgent issue can have a life of its own, quite apart from the issue of scientific consensus. When public health regulators come under pressure from the general public — as they are today on the subject of EMFs — they've been known to take action on the basis of worrisome, albeit incomplete, data. In fact, this happens more often than not. In most cases, regulators have to take other things besides scientific proof into consideration, which is probably just as well, because if we were forced to wait for scientific consensus before we could act to protect people, it would only slow down an already cumbersome regulatory process.

The question of when to take regulatory action isn't a simple issue of scientific proof. It's fraught with the same economics and politics of other governmental decisions. In the case of EMF health risks, the people have been lied to repeatedly, the scientific truth has often been obscured as studies are hidden and their results are denied, and public officials and experts who have looked to their consciences and tried to warn the public have been gagged or subjected to attacks in an effort to discredit them.

Interestingly, in this age of "techno-reason," one of the things that is often factored into a public health decision is emotion or how the man-on-the-street feels about the issue, especially how much it worries or frightens him or, as Dr. Peter M. Sandman, director of the Environmental Communication Research Program at Rutgers University, would say, "outrages" him. What the people themselves want can and does influence regulatory decisions.

In his 1990 speech about EMFs, M. Granger Morgan, an expert in risk assessment, said, "Risk assessment poses great challenges. Public perceptions are likely to play a major, not a minor, role." He went on to report that

What The Studies Tell Us

there was a growing public pressure for EMF standards and that utilities and public health organizations were going to have to think about the strategies they should adopt, warning, "Incremental extrapolations from business-as-usual may not be adequate."

How much proof does it take before an agent is considered hazardous? As a society, we often act when there's even a possibility of very minuscule risks. Take the EPA regulation of pesticides: One study warning that a pesticide poses a potential threat to health is enough for the agency to ban that chemical — in fact, the EPA is required to do so. For instance, in 1986, the EPA received a report that rabbits exposed to a particular defoliant developed severe defects, and the agency acted immediately to ban the chemical. A drug that was found to cause cancer in test animals, as EMFs have, would be taken off the market — the assumption being that it could also cause cancer in humans.

We often act to regulate when data are limited, but of grave concern. And what could be of greater concern than cancer from daily exposure to the electric power system, especially when some of the most compelling data about the risk involve children? Each year that we wait, based on the most conservative estimates from the early epidemiological studies, about 13,000 more children will die of cancer because they've been exposed to power-line magnetic fields; not to mention the electrical workers, the electricians, the TV and radio repairpeople, and people like you and me.

The National Cancer Institute reported a twenty-eight percent increase in childhood cancer from the fifties to the eighties, but little is being done to address the problem. As a society that professes to care greatly about our children, how can we continue to just stand by when study after study informs us that at least twenty percent of the increase in childhood cancer in the United States is being caused by our electric power system?

Furthermore, all those studies of children aren't simply about children — they're about all of us. Remember those canaries that coal miners used to carry with them down into the mines, to warn them of a sudden dangerous rise in the level of carbon monoxide? Well, our children are like those canaries. Their deaths are a warning that adults are also in danger from power-line cancers. Why are the data more clear on the children? They are more sensitive to toxins (possibly due to the fact that their bodies are still developing, so their cells are involved in the process of mitosis). Second, they make easy research subjects because they're not very mobile and not likely to be subjected to the large variety of toxins that adults are subject to. Also, many children diagnosed with cancer live tragically short lives, so it's easier to collect mortality data on them than on adult cancer victims.

Risk assessment, or risk management, is the business of managing health risks. As with any other discipline, there are various schools of thought in public health about how and when to define a health risk. But, in situations where the probability of risk seems reasonably pervasive — like the case of involuntary exposure to ambient magnetic fields, since every one of us comes into daily contact with the agent — many experts wouldn't insist on scientific consensus at all. Instead of waiting for more information, once a potential health hazard of this magnitude has been identified, they would prefer to get to work warning the public to take out an insurance policy by establishing a regulatory safety net to prevent what could emerge as a major public health disaster.

Where EMFs are concerned, most public health people will at least counsel a prudent avoidance approach, meaning that people should do what they can to limit their exposure without spending a lot of money or greatly changing their lifestyle. Clyde Murley, of the San Francisco Public Utilities Commission, believes the time has come for EMF regulatory action. Stating that he's "very concerned about 8,000 more kids dying of cancer every year," he went on to explain some of the problems with regulating EMF exposures:

> Typically, what society does is regulate what is easy to study. But, if someone loses his keys at night, that doesn't mean they'll find them under the street light. I think difficulty of testability should be factored in along with the usual problem-times-numbers-equals-risk equation. We should consider how difficult it is to study a particular scenario when we're making decisions about regulating it. For instance, it is very difficult to study EMFs and cancer in the lab, but far easier to study it epidemiologically. This kind of resistance to testability really hampers regulation.

Chapter 5

EMFs in Your Community

Our utility concerns are concerns about being able to manage magnetic fields from our lines if that should be proved to be necessary. At present, there's no interest throughout the industry in putting these techniques into effect. We'll leave that up to the decision makers, society at large, who will have to decide in the end whether the gain is worth the expense or the inconvenience.

> LUCIANO ZAFANELLA,
> manager of the EPRI
> High-Voltage Transmission
> Research Facility

The possible health risks of electromagnetic fields are fast becoming an urgent social issue. Numerous public meetings and forums around the country have been called to discuss EMFs to help people better understand the subject of electromagnetic fields in general and EMF health risks in particular, or to cope with specific EMF problems in their communities, like the siting of a new power line or the discovery of a cancer cluster in the vicinity of an electric substation. Despite the complexity of the topic, however, public meetings about EMFs tend to be remarkably similar.

The people who come to them are worried and usually confused. They want to hear what the panel of experts has to say about EMF health risks. They expect the experts to clarify the issue for them. Unfortunately, these panels are usually made up of such a mixed bag of "experts" — skeptics, believers, scientists, environmentalists, special-interest group spokespeople — that they often add to the confusion. Each panelist will proceed to give his or her side of the issue, which usually conflicts with someone else's views. Somehow, the public has to sort it out.

WARNING: the electricity around you may be hazardous to your health

During the latter part of the program, people in the audience are invited to get up and address questions to the panelists. Wherever the meeting is, whatever the audience's agenda may be, people always seem to ask the same questions. Because of the similarity of these meetings, a public information meeting on electromagnetic fields called by the EPA and the City of New York in the fall of 1991 will serve as an example.

The meeting was held at New York University. The panel of experts included David Carpenter (dean of the State University of New York School of Public Health), Louis Slesin (editor of Microwave News), Indira Nair (Carnegie Mellon University), Herb Kaufman (ESEERCO — Empire State Electric Energy Research Corporation), John Wilson (Consolidated Edison), Stan Sussman (EPRI — Electric Power Research Institute), Martin Blank (Columbia University), Michael Buccigrossi (U.S. EPA), and Robert Kulikowski (Bureau of Radiation Control, New York City Department of Public Health).

The public, some 150 strong, looked remarkably like a cross section of middle America. When it came time for the people to have their say, it was apparent that they had done their homework: They understood the subject well enough to ask some very specific questions, and they had no trouble at all articulating their concerns.

First, each of the panelists gave a brief presentation on the EMF/health issue. Then the people got to query the experts. Questions from people seeking general information were fielded by various members of the panel, according to their bias. Often, other panelists representing a differing perspective would then respond.

But many of the queries that night went beyond the level of general information. Whereas the scientific community might still be withholding their opinion, the majority of the people who are informed about magnetic fields seem to be convinced there is a problem. This was clear at the New York meeting. People wanted to know two things: how to reduce their exposure and who had the responsibility to protect them from EMFs. When panelists informed the audience that no one agency or utility currently has that responsibility — an official from Consolidated Edison told a speaker "you're on your own" — the crowd broke out in boos and hisses. About halfway into the question and answer period, the crowd seemed to lose patience, and the evening took on an adversarial air. As scenario after scenario was presented, it became apparent that the Con Ed man was right. No agency or utility was willing to even come out and measure the strength of magnetic fields, much less suggest some amelioration. For the most part, the government agencies and the utility spokespeople denied there was any danger at all.

Carpenter, Slesin, and Indira Nair, however, were advocates on behalf of the public. At the start of the meeting, Carpenter gave an impassioned plea for people to "speak out" and demand some protection. His comments were seconded by Slesin, who spends half his time at meetings like this trying to keep the record straight.

Dr. Indira Nair is a world-acclaimed physicist in Carnegie Mellon's Department of Engineering and Public Policy and is co-author of the 1989 Office of Technology Assessment paper that redirected the scientific community's attitude about EMF bioeffects with the statement, "The emerging evidence no longer allows one to categorically assert that there are no risks." Eventually, Dr. Nair became quietly vociferous as she picked up on the public's frustrations. At one point, she produced a slide that she had held back during her initial statement, introducing it by saying,

These are my concerns (about EMF exposures), not Carnegie Mellon's:

> (1) We're changing the evolutionary environment. (2) Certain types of cancer have shown an increase. (3) Worries about the central nervous system effects. . . . The preponderance of the evidence tells me something is there I should worry about.

Dr. Nair is one of the team who coined the term prudent avoidance: People should do whatever they can to avoid exposure to magnetic fields if they can do so without great cost or great inconvenience. This has become a device the experts have used to pass the problem back to the citizens without really giving them any help. Late in the evening, people got tired of being told to exercise prudent avoidance. A man in the audience said, "They're building a substation forty feet away from an elementary school in our neighborhood. After it's built there, how are those 800 kids going to practice prudent avoidance?"

Dr. Nair responded by saying, "We certainly didn't intend prudent avoidance to be used in relation to substations, transformers, high-voltage lines." Her answer to people's concerns was very direct: "If the fields near your house are much higher than normal ambient fields, move away."

Louis Slesin also "got more and more frustrated" as the night wore on. He finally told a man, "Look. The reason no one wants to give you an answer is if they do give you an answer, they'll be in a lot of hot water. The problem is, we don't really know if we have confidence in the 2- to 3-mG level. And if it's a problem for you, then it's a problem for most Americans as well."

Most of the people at that meeting shared Slesin's worry. They had come for one reason: They had a problem with electromagnetic fields and they wanted help. Like you, they want to know whether they're in some danger from exposure to magnetic fields and what they can do about it.

WARNING: the electricity around you may be hazardous to your health

THE MOST COMMON QUESTIONS RAISED IN EMF FORUMS

- Should I buy a house that's 300 feet from a 500-kV transmission line? Will I be able to resell it or is it going to go down in value because of its proximity to the line?
- We have two children in elementary school. We recently had our home measured and the magnetic fields in our son's room were generally in the 20-mG range. What should we do?
- Our son's school has a primary distribution line going right over it. Is that dangerous to the kids? How can we get the utility to move the lines?
- Who can I get to come out and measure the magnetic fields in my apartment?
- Do EMFs produced by vehicles affect you?
- Is anyone doing anything about regulating the siting of new power lines? What about a moratorium for five years while they figure out more about the risks?
- What about all these new substations around the city? Are they dangerous?
- What agency will come out and measure the fields on my property? I tried calling the electric company and the Department of Public Health, but they said they don't do it.
- How can I practice prudent avoidance when they're building a substation cheek-to-jowl with my building and with the sidewalk, and they've measured fields of 128-mG at the outside wall? And what about the feeder lines? Aren't they dangerous?
- Who can I get to measure the fields from the substation that operates the Long Island Railroad that's across the street from us?
- Why is Con Ed expanding all their substations around the city when there might be a health hazard? How can they claim there's no hazard with all these studies? Who has the responsibility here?
- Where can I get a meter? How do I know what to measure? Should I get a broad-band measurement or just 60-Hz? What about harmonics and things like that? Are there generally more things going on around us than I've ever heard of?
- In our neighborhood they're putting in a substation just forty feet from an elementary school. After what we know, can't the town stop them? What do we have to do?
- Are cable TV lines dangerous?
- What about the power source for those cellular phones? Are they dangerous? They put one in my building, and when I asked about it, no one would talk to me.

EMFs in Your Community

WHERE TO LOOK

It may be that you and your neighbors have already identified an EMF problem in your community. Maybe it's a mysterious cancer cluster. Or it might be one of those common power-line siting issues, perhaps an existing power line that you now think is dangerously close to residences or a proposed new power line whose route you feel will threaten people's health. If an issue of this sort has already been raised in your community, you may have begun to make your concerns known to your elected officials, community boards, regulatory agencies, or the power company itself. If so, the discussion of effective community organizing strategies at the end of this chapter, along with some of the models throughout the book, will be helpful.

On the other hand, you may have only recently had your consciousness raised about the hazards of electromagnetic fields and you want to find out whether or not you and your family are being exposed to dangerous fields. To do that, you need to take a physical inventory of the electric landscape around you, outdoors in your own community. Look for components of the electric power system that are close to populated areas, particularly:

- Homes or schools or public recreational facilities, especially children's playgrounds, that are less than 350 feet from high-voltage transmission lines. Playgrounds and schools located near electrical components with high magnetic fields are of particular concern because the developing child is in more danger of cancer from magnetic field exposure than an adult would be. Far too many children's facilities are sited near, or even directly under, high-voltage lines. Existing right-of-ways in this country were established to protect the public from the danger of shock hazards, not magnetic fields. Therefore, communities saw no problem with siting schools and parks on right-of-ways where the land was public property, and as such, inexpensive and easy to zone.
- Homes or schools or playgrounds transversed by high-voltage transmission lines or high-current distribution lines. Remember, during peak times, the magnetic fields under distribution lines are often as high as those emitted by high-voltage transmission lines — and that the epi-studies have linked increased childhood cancers to distribution lines.
- Homes or schools within fifty feet of primary feeder distribution lines (see Figure 2-A, in Chapter 2, for help identifying them). Distribution lines are almost always on wooden poles about thirty feet from the ground, and transmission lines are mounted on higher, metal towers.
- Homes or schools or playgrounds less than forty feet from pole-mounted step-down transformers.

How high are transmission-line fields/ What do transmission-line magnetic fields measure? According to Marvin Chatkoff, coordinator of the electrical engineering program and the University of Texas, San Antonio, the following field strengths are typical:

- 750 feet away from center of right-of-way for two 345-kV lines, the magnetic field would be 5 mG.
- 800 feet from a 756-kV line, the field would be 5 mG.

Children's exposures are of special concern because so much research has linked EMFs to increased childhood cancer. Children's exposure patterns are generally easier to assess than adults' because they're less mobile. Since they spend so much time at school, it's essential to assess the magnetic fields in their classrooms as well as those at home. And while you're inside the school, don't forget to measure the computer room, fields from fluorescent lighting, and playground areas near electric power transformers or lines. Also measure the playgrounds and ball fields the children frequent after school, especially if any of these are near suspect high-current electric power lines or substations, as is often the case. It's important to measure the fields where they wait for the school bus and, if they walk to school, you should go over their route yourself and check for high fields. And don't forget the places they hang out after school. (You may want to declare video arcades off limits immediately.)

But do all this measuring, if possible, when the kids aren't around, to avoid scaring them. It's a good idea to sit down with them and explain the problem, especially if they've heard about magnetic fields and know that you're concerned. (And, so far as their at-home exposure is concerned, if they're old enough to use appliances, they need to be educated about keeping their distance from things like the TV and the microwave oven, as you'll see in the next chapter.) Try to do all this in a reassuring and somewhat understated way.

School districts around the country are taking the initiative and having their EMFs measured. The Martha's Vineyard School District hired Karl Riley, whose company, ELF Magnetic Surveys, specializes in locating the sources of high magnetic fields. Riley says,

> In the Edgartown School, I found that kids in one of the classrooms were sitting in a 30-mG field. There was a wiring mistake in one of the subpanels. They got in an electrician — they were able to find one who was interested in doing this sort of thing. Although the problem was new to him, in short time he was able to understand what we were attempting to achieve by

putting the hot and the neutral together to result in balanced magnetic fields. So, in about half an hour, he was able to rectify fifty years of exposure.

In 1994, the California Department of Health Services began a $7 million research and educational EMF program. This extensive project, which is due to report in 2000, set out to identify potential EMF hazards and make reasonable cost-effective recommendations to mitigate them. An important component of the program is the California EMF in Schools Task Force.

Karl Riley has been active in this program as well; measuring EMFs in schools all over California and mitigating them as part of the research effort. Riley says he found that elevated magnetic fields were almost always caused by improper interior wiring that had resulted in net current in the wires — not by the power lines outside a school. "Internal wiring never gives a net current unless there's a wiring error," Riley explained. "And this can always be repaired." When it's repaired there's no field.

Currently Riley is also carrying out what he describes as "the largest MF mitigation job anyone has ever done," repairing wiring errors in thirty buildings on the Grossmont College campus in El Cajon, California. Riley did the initial EMF survey back in 1995; the college got a federal grant to do the cleanup. "Once they decided to go ahead and do this, they've really been a model for this sort of project," Riley said.

The EMF program has put together a very helpful "EMF Checklist for School Buildings and Grounds Construction" that school districts can use to limit EMFs when siting new schools. The checklist is full of what its authors call "low- and no-cost techniques" for EMF mitigation. (Interestingly, the DOE refused to distribute it.) This excellent resource can be obtained for $6 from Copy Central, 5801 Christie Ave., Emoryville CA 94608.)

You'll often find that property where power lines are sited has been designated open space or turned into a park or a par course. Many health-conscious people are taking their daily jog in dangerously high magnetic fields; if you're working out in a power line right-of-way, a word to the wise is sufficient.

Dr. Fred Stertzer, ex-chairman of the Nonionizing Radiation Advisory Committee of the New Jersey Commission on Radiation Protection, feels very strongly that this practice of siting children's facilities near power lines must be stopped. In fact, the New Jersey commission was one of the first bodies in the country to develop a rule on this EMF hazard. "Years ago, we suggested simply that people not put playgrounds under lines," Dr. Stertzer explained.

WARNING: the electricity around you may be hazardous to your health

Dr. Stertzer says the issue of health risks of EMFs is more and more a subject of public concern. "Recently there have been a lot of citizens' complaints that power companies would run lines near residences or playgrounds or schools. I'd say there's been a fair amount of agitation on this. People are upset."

The same is true in other countries as well. In 1990, in response to citizens' protests about the siting of a high-power transmission line, the Soviet Union ordered the utility company to make the line direct current. The Swedish National Energy Administration recently issued an advisory warning that schools, playgrounds, and day-care centers should not be sited near power lines. Sweden is also ahead of most other countries in setting up magnetic field regulations. The Swedish standard extrapolates from the epidemiological research, recommending that children not be subjected to magnetic field levels over 3 mG. (In the absence of any clear guidance from the U.S. government, many Americans have chosen to adopt the Swedish standard.)

Dr. Daniel Wartenberg, the present chairman of the New Jersey Radiation Commission, has called for a fifty percent reduction in magnetic fields from all new power lines over 100 kV. He points out that this will mean an additional fifty percent in construction costs, less than one percent of the line's lifetime operating costs. These days, many states like New Jersey are concentrating EMF regulatory efforts on the siting of new power lines.

In the mid-nineties, Sweden's Energy Administration recommended a moratorium on the siting of new houses near high-voltage transmission lines "until more conclusive research is available."

The problem of new homes being built too close to high-voltage lines is a very real one in the United States, where so many communities have experienced rampant growth in the past decade that open land is at a premium. Because of the absence of right-of-way rules regarding the hazards of magnetic fields, today we see thousands and thousands of homes going up within 300 feet of high-voltage lines.

In Daly City, California, the planning commission allowed a developer to build dozens of expensive houses directly beneath two sets of high-voltage lines. These $350,000 homes are also right across the street from a large transformer yard that has been identified as a hazardous waste site because of its high level of PCPs. Prospective home buyers are required to sign a disclosure statement that reads, in part:

> The subject property is located near Pacific Gas and Electric and the City of San Francisco high-voltage electric transmission lines. Purchasers should be aware that there is ongoing research on possible potential adverse health effects caused by the expo-

sure to electric and/or magnetic fields generated by high-voltage lines. Although much more research is needed before the question of whether electric and/or magnetic fields actually cause adverse health effects can be resolved, there is a possibility that such a risk exists. Residents with pacemakers could be adversely affected. ... At this time, no risk assessments have been made.

The statement goes on to warn that the homes are on the three major active faults in the San Francisco area, "the epicenters of all major, and most of the minor, earthquakes in the Bay Area," as well as a large hazardous waste site. Nonetheless, these houses are selling like hotcakes.

San Francisco Realtor John Barry was concerned that people didn't understand the hazard. He wrote to the State Department of Health Services in January, 1990, suggesting they consider this as a statewide health issue:

> Many of the buyers of these homes will not have their own agent or lawyer and the (disclosure) will be glossed over, if discussed at all. Many buyers don't speak English. Is it right to transfer the risk to them in this way? You might involve the State Department of Real Estate to augment its disclosure law so that the buyer is forced to have some sort of informed opinion given to them by some neutral agency, other than the (environmental impact) report written by the company hired by the developer.

Besides the obvious high-voltage lines and substations, there are additional sources of high 60-Hz magnetic fields. In most communities, overhead distribution-line fields are more dangerous to the average person because most people generally don't come in contact with transmission line fields. Overhead distribution lines are everywhere, on every street, exposing people to fields that can go up to 20 mG during peak current times. But, since they're always strung on only one side of the street, you can avoid these fields completely by crossing the street and walking on the side without the lines. Burying lines can be very effective in reducing magnetic fields. Cables are wrapped together, thereby canceling out the fields. Often, strong magnetic fields underground are coming from current on water pipes.

Factories and large office buildings have their own transformers, usually buried outside them or housed in a shed on the property. Large high-rise apartment buildings in many cities have transformers on the premises to step up and step down current, as well as high-current wiring throughout that may be dangerous, because wiring is the purview of the landlord rather than the public utility.

WARNING: the electricity around you may be hazardous to your health

Subways and electric trains or buses also produce high magnetic fields, both on and around them. Joseph Bowman, of NIOSH (National Institute for Occupational Safety and Health), said he's measured very high fields in San Francisco apartments on streets where the electric buses run. These electric transportation systems produce very high magnetic fields around them at stations, turnarounds, switchyards, and wherever their transformers are located. And people who depend on this electric public transportation are riding back and forth to work every day in extremely high magnetic fields.

Lynne Gillette and Doreen Hill of the Environmental Protection Agency's EMF group conducted an informal study of colleagues' ambient magnetic field exposures by supplying them with dosimeters and having them keep a log. Exposure periods were divided into home, work, transportation, outside (other than work), and inside (other than work). They presented their results to the Bioelectromagnetics Society at a meeting during the summer of 1991. Surprisingly, the people — all working adults in a major metropolitan area — got most of their exposure, as well as exposure to the highest fields, outside of their homes and offices, during transport. Gillette herself takes Amtrak "five days a week, one hour each way." She got readings of 500 mG on the train. Gillette believes that the reason spot measurements in many studies don't correlate with increased risk is, "They just don't correlate with exposures any real people are getting. One-time spot measurements in the home simply don't work." (In fact this may have been a problem with a number of studies that used one-time spot measurements and reported no relationship between EMFs and cancer.)

Besides exposures from 60-Hz power-line fields, people in every community are also regularly exposed to radio frequency (RF) and microwave (MW) radiation. These come from radio and television broadcast transmitters; military acquisition and tracking radar; civilian and military air traffic control systems; and weather radars; telephone microwave relay systems and other wireless communication transmissions; satellite communications earth stations (SATCOMS); ham microwave radios and mobile CB radios (See Chapter 8: RF/MW Radiation.)

MEASURING EMF LEVELS AROUND YOU

To ascertain the actual levels of the magnetic fields around you, they must be measured with a magnetic field meter called a gaussmeter. You can obtain a field measurement by hiring a professional or you can buy a meter and do it yourself, which is a cinch with the proper instrument. (See Appendix D for listings of testers and companies that offer affordable gaussmeters.) You can also rent gaussmeters from engineering firms in many cities these days. The main difference between the two solutions is cost. It will probably cost less for you to purchase a meter and take readings yourself — the going rate for such services is about $300 a day, including a report. Once you own a meter, you'll be able to whip it out and measure fields wherever you are. On the other hand, a professional can offer you not only technical expertise but a written report, which will come in handy if you have to deal with agencies or boards, plus the added benefit of useful advice derived from experience. Sometimes it really helps to talk to someone who's able to put your own exposure scenario into a larger framework of exposures seen in other places.

David Bierman is a self-proclaimed "house doctor" with Safe-Environments Home and Office Testing Laboratories in Berkeley, California. Bierman says he's spending more and more of his time lately measuring magnetic fields for homeowners. In particular, a lot of people want to know the level of fields before they purchase a new home. In addition to conducting field measurements on a variety of the locales featured in this book, Bierman is a wealth of information about EMF scenarios brewing around the country. Bierman is a member of the National Electromagnetic Field Testing Association in Evanston, Illinois.

If you decide you want professional help to measure suspicious fields, but you don't want to have to pay for it, try calling your electric utility. A few utility companies will come out and measure fields on your property (Most of them won't do this for tenants, however.) According to company spokesperson Catherine Moore, Pacific Gas and Electric in San Francisco "will on request measure EMF levels in neighborhoods and homes, but it cannot interpret the numbers." And be warned that electric companies are notorious for coming around at times when the fields are at their lowest. (Magnetic fields are a function of the current in the lines, so they can fluctuate greatly and should always be measured at peak current times to obtain highest measurements.)

Many electrical engineering firms these days can also provide this service or will refer you to someone who will. With the climate of concern right now, hardly a week passes without another environmental firm advertising magnetic field measurement services. Look for ads in newspapers and magazines.

There are also some very good EMF consultants around the country — like Karl Riley or Cindy Sage of Sage Associates in Montecito, California — who can assess fields in your community and help you with mitigation. Some private companies even can be hired to bury power lines. They offer design, engineering, and installation services, at prices that are competitive with the utilities. Burying power lines makes particularly good economic sense in new developments, since setback requirements can be reduced and

WARNING: the electricity around you may be hazardous to your health

density increased. Homes in developments where power lines are underground also command higher prices.

According to Sage, "Undergrounding is being widely done in Europe and Japan and it's becoming more prevalent in the U.S. It can reduce EMF and eliminate visual impact from transmission lines. Cross-linked polyethylene is a newer product option for voltages up to 135 kV and has significant cost savings in comparison to traditional oil-filled pipe type cable. At the lower volts, private contractors can install low-pressure fluid-filled cable and still be very competitive with utility pricing."

If you decide to purchase a gaussmeter and do the measuring yourself, there are a great variety of designs ranging from small personal gaussmeters, called dosimeters, that can be worn on the body to obtain twenty-four-hour exposures, to meters that cost thousands of dollars and take broad-band readings all the way up to the microwave frequencies, to meters that print out the data for you as you go along. The inexpensive survey-type meters are hand-held, portable instruments that give a digital readout in milligauss as you go along. They are extremely simple to operate.

Basically, gaussmeters are of two types: two-part meters that consist of a separate sensor (a coil-like device that picks up the field) and a meter body that contains the readout device, or designs that combine both parts in one module. For the most part, the less expensive types designed for the average homeowner are small, light, and easy to use. My Magnetic Sciences International (MSI) MAG CHECK 20/25 has a separate sensor/probe that can be plugged right into the meter body for easy handling or can be connected with a long wire to allow readings at a distance.

Before you conduct your survey, be aware that you may find yourself standing in some very high electromagnetic fields. Always take care to avoid electric shock from the electrical sources of the fields you're trying to measure. (Meters themselves are designed to minimize shock hazard, of course.) Also take into consideration the potential hazards of the magnetic fields you're subjecting yourself to and be prudent about limiting your actual exposure time.

With all this in mind, you're finally ready to measure those fields. First, read the printed instructions that came with your meter. They should be detailed and easy to follow. Once you've followed the directions in your manual and set up your gaussmeter, you're ready to "take action to diminish your exposure and the exposure of your family and friends to magnetic fields," as Magnetic Sciences International, the manufacturer of my MAG CHECK meter, puts it.

The main thing you have to do is get the feel of how much you have to rotate your meter — or, if it's a two-part model, the sensor — to get an accu-

EMFs in Your Community

rate reading. The idea is to continue rotating it until you get the consistently highest reading. If you have a two-part gaussmeter, it may take a little practice to slowly rotate the sensor while you keep the other component as still as possible so it can settle into an accurate reading.

You can also ascertain the levels of power-line fields around you by using standard computation methods widely employed by the industry. In fact, electrical engineers working for utilities hardly ever go into the field to measure EMFs — they sit at computers with special programs designed to do that for them based on the voltage on the lines. Or they use publications like the Transmission Line Handbook, issued by EPRI, that contain physics equations for electromagnetic fields near power lines.

An additional method that has gained acceptance was used in some of the epidemiological studies to assess the level of fields based on wiring configuration codes. These codes were developed to estimate current and hence magnetic field strength based on the lines near a house. (Chapter 6 describes these wiring configurations in detail and shows how to use them to assess the fields in your home.)

EXPOSURE LIMITS

After you've gone outside and looked for the sources of dangerous magnetic fields in your community, what do you do next? Let's say you spotted something in the electric landscape that raised a yellow flag, perhaps a high-voltage line near a subdivision or a substation at the edge of a school playground. Once you've identified a possible source of magnetic fields, your immediate task is to obtain a measurement of the actual fields to determine whether or not they could be making you and your neighbors ill.

There's a heated debate on what levels of EMF to consider dangerous. Since the experts haven't come to a consensus on this, we each have to decide for ourselves. Many of the experts, myself included, think people should stay out of magnetic fields of 2.5 mG and higher; in many, many studies, this is the point where adverse health effects begin. NCRP recommends 2.5 mG as their safety limit. Remember that, at some future time, lower levels may be determined to be harmful and you will want to adjust this safety level downward.

For the most part, where there are magnetic field problems, readings are almost always much higher than 3 mG. They're likely to be in the neighborhood of 6 or 10 or even 20 mG and above.

As a rule of thumb, consider what I call the three D's of exposure: Dose, Duration, and Distance. Dose refers to the level of the field (over 3 mG is to be avoided); another concern is how long you've been in the field, the

WARNING: the electricity around you may be hazardous to your health

Duration of the exposure; and your Distance from the source tells something about how hazardous it is, since magnetic fields drop off as you increase your distance from the source.

The idea that more is worse and the longer the exposure, the more worrisome the exposure, makes sense. This is the model generally applied to carcinogens and other toxins and — until someone proves otherwise about EMFs — I will subscribe to it. Even if it turns out that the health hazard is linked to some particular component of the magnetic field, as some think, like windows or transcients or pulsed waves — if you stay away from high fields, you'll protect yourself from each of these.

The question still remains: Once you identify those high magnetic fields as potential health risks to people in your community, what should you do? It will really depend on how much you want to see change and how willing you are to persevere.

As you've seen throughout the book, once you've become aware of a magnetic field hazard, you have to call for some sort of community forum to inform your neighbors and your elected officials about your concerns. This may also be the opportunity to begin a dialogue with the electric company.

With this book as a reference, you and a group of other concerned citizens should be able to put together a convincing fact sheet to distribute to others who aren't aware of the potential dangers of magnetic fields. Then try to get some reliable experts to come and talk to your community. This initial organizing step can be a time-consuming process, but no effective lobbying for change can occur unless you have your community behind you. Until people believe there's a real problem, no one's going to do much about trying to change it. You'll be doing a real service by raising the awareness of home buyers in your community to dangerous EMF sources around you.

At the start, at least, try to avoid taking an adversary position until one is forced on you, and do your best to avoid polarizing your community. Have the attitude that the school board or the city council or other property owners are just as concerned about public health as you are. When you do reach a consensus of concern in your community, what sort of recommendations should you make? That will depend on the particular problem you're trying to address. But to give you a few ideas about possible solutions, here are some common routes we've seen communities take.

Let's say the issue revolves around the siting of a new power line. In that case, you should address your elected officials and bring the problem before the Public Utilities Commission (the Public Services Commission, in some states) with as much backing as you can muster from the community.

The situation with right-of-way (ROW) regulations of magnetic fields cries out for attention. Some communities have succeeded in widening existing right-of-ways. Of course, this only works when there's still open land around the lines.

Most existing ROW regulations don't even take EMFs into consideration; they were designed to protect people from the danger of electric shock and that's still all they do. They don't address the problem of EMFs. As one utility spokesperson put it: "Absent any good information, regulations aren't necessary." The only thing that will have any effect on the industry's stonewalling about this is citizen activism at a grass roots level. This is where your campaign to protect people from EMFs in your community can have some effect, especially if you take time to educate people from the start.

Existing ROW rules generally allow homes to be built within 300 feet of high-voltage lines. This is hardly protection from EMFs, when you consider that Joe Norton, transmission-line coordinator for the state of Florida, has measured 6-mG magnetic fields 2,000 feet away from a 500-kV line. Norton, who has spent years in the thick of the EMF standards battle in Florida, is not optimistic about the future of such regulation. "The reason why regulations aren't going to be set is liability," he explains. "No government body wants to be ahead of anyone else on this issue. The nature of government is to stick to the status quo." (By the way, the law firm that represents the utilities has written the procedure for power-line siting in Florida.) The Department of Environmental Regulation is dragging its feet on this, to the point where Hillsborough County is considering reopening its EMF lawsuit to force the state to set safe EMF limits.

Some states, among them Florida, California, and New York, have tried to establish interim limits to protect their citizens from EMFs — but most of these limits are too high. For instance, New York sets magnetic field levels at:

- 150 mG at the other boundary of a 230-kV or smaller transmission line right-of-way or at the property boundary of a new substation.
- 200 mG for the maximum magnetic field at the outer boundary of the transmission line right-of-way for a 500-kV line or at the property boundary of a new substation serving a 500-kV line.

California's State Department of Education established guidelines to keep high-voltage lines away from schools in the mid-nineties:

- 100 feet from easement for 100-110-kV lines
- 150 feet from easement for 220-230-kV lines
- 250 feet from easement for 345-kV lines
- 350 feet from easement for 500-kV lines

Unfortunately, these recommendations may not be stringent enough. The department takes a conservative approach in reviewing school sites that are to be situated near power transmission-line easements. It sets forth other factors that must be considered when reviewing site plans where a school is situated near a power-line easement, such as:

1. Why is it necessary for the district to build the site near the easement? What other options are available?
2. Has the district contacted and discussed with the utility company whether the company intends to increase the voltage of the transmission lines?
3. Does the utility company intend to build other towers on the easement at any time in the future?

Another item on the agenda for communities intent on limiting people's magnetic field exposure is regulation of the routes of new power lines. The rules, of course, should be simple: Keep them away from heavily populated areas and children's facilities.

Try to give the utility the benefit of the doubt. For one thing, the magnetic field exposure citizens are experiencing may not be entirely the company's fault. It may be that in your particular community, as in many around the country, growth has encroached on power lines that were once far away from people. And it's a fact that today, more and more private power companies around the country are taking responsibility to protect the public from high magnetic fields from their lines. It may surprise you to learn that a number of utility conglomerates are advising their members to site new lines away from densely populated areas.

According to environmental planner Andrew O. Linehan, utilities are being forced to take into account electromagnetic fields as a regular part of siting decisions. In a paper he presented to the Washington State EMF Task Force in September, 1991, Linehan stated: "The EMF issue has evolved to the point where communities and regulators expect utilities' route selection studies to explicitly consider EMF avoidance as an alignment evaluation factor." EMF avoidance can be reflected in siting criteria by: "Giving high priority in the corridor selection process to avoiding existing and proposed residential, school, and other high-use areas" and identifying "alignments that minimize exposure of sensitive receptors to EMF." (Sensitive receptors include houses, schools, parks, and office buildings.)

Engineers working for utilities are hard at work developing ways to mitigate magnetic fields from their lines. For one thing, they've developed new designs for distribution systems that include low-voltage secondary ground-

EMFs in Your Community

ing procedures to reduce magnetic fields. They're also trying, whenever possible, to balance lines — grouping them together so the charge on them is generally neutral, which eliminates high magnetic fields. New power-line configurations can cut magnetic field emissions in half.

It's a lot more difficult, for a variety of complicated reasons, to balance distribution lines, however. According to Professor Stewart Maurer, project director for the mitigation research project being conducted by ESEERCO, "In the case of a distribution line, the only solution is to have it as far as possible from a residence." (We'll talk more about distribution-line fields in Chapter 6.)

The power companies have already spent a great deal of money to develop a variety of engineering methods to manage power-line magnetic fields. Some of these techniques are:

- Creating a low-voltage countercurrent to make the overall current on the line closer to zero. (One line is high-voltage, the other is low-voltage, and they're placed closer together so they cancel each other out.)
- Undergrounding, or burying, lines closely grouped together.
- Using special circuit configurations that reduce fields, such as the delta configuration, may become a possibility. There are problems with this solution because it would require replacing every piece of electrical equipment on old lines and thus would be very costly. (This system is currently used in Sweden.)
- Spacing conductors (lines) closer together, called compaction, or bundling. This can also increase the electric shock hazard for workers. (One of the things engineers are working on is the development of robots that can take the place of humans in line maintenance where there is an increased danger of shock.) Bundling presents another problem: The closer together conductors are, the more the possibility of snow or ice building up on them and causing power outages.
- Using a different arrangement of the three phases in the lines so that the phases more closely cancel out one another.
- Using a low-reactance reverse-phase configuration in certain circuits to cut magnetic fields nearly in half. This is a relatively inexpensive solution, especially when used for new lines, and it is currently in operation in a few parts of the country.
- Changing phases right at the substation so that they balance more and the magnetic fields are reduced.
- Controlling the amount of current on a line by changing the number of lines and the voltage on each one. For example, instead of one high-voltage line delivering 40,000 to a street, three lower-voltage lines could provide the ser-

vice. However, this could still be a problem in peak times when the demand for high power from lower-voltage lines might actually produce more current than was experienced with the higher-voltage distribution system.
- Using six- or even higher-phase conductors instead of the usual three-phase. Called high-phase-order (HPO) transmission, this method would enable distribution lines to carry high amounts of voltage while cutting down on fields. HPO transmission is used in many countries for residential power systems, but in the United States it's only used to deliver electricity to factories.
- Placing lines closer together in existing right-of-ways instead of building lines through new routes. The additional lines could be designed to cancel out the original line currents.
- Modifying configurations and using new materials and new circuitry designs or configurations to produce lines with lower fields.

These mitigation methods have their pros and cons, and it's not likely they will be widely used. In fact, according to Luciano Zafanella, manager of EPRI's High-Voltage Transmission Research Center in Massachusetts (a model community that has been set up for field management research):

> At present, there's no interest in putting these techniques into effect. Our ultimate concerns are concerns about being able to manage magnetic fields from our lines if that should be proved to be necessary. Any changes in the power system imply changing characteristics of the electric system that could cause problems in providing service. We'll leave that up to the decision makers, society at large, who will have to decide in the end whether the gain is worth the expenses or the inconvenience.

In the case of power-line magnetic fields, remember that, by and large, the only one who can practice prudent avoidance is the electric company itself. In the interests of public protection, the utility and your elected officials should be able and willing to thrash out some sort of solution to EMF problems in your community — from fencing off an area on a school playground, to redesigning the circuitry in a row of pole-mounted transformers that come very close to an office building, to simply declaring a moratorium on the siting of power lines that produce high magnetic fields near heavily populated areas.

In '97, when the Swedish government recommended that electric companies use prudent avoidance when siting new facilities "to reduce the risk of injury to human beings," they also recommended reduction of EMFs on existing lines — when this could be done at a reasonable cost. The cost of mitigation certainly has to be addressed. A handful of research projects are assaying the cost of large-scale EMF mitigation and weighing it against what many think of as a relatively small public health benefit. Rather than undertake the enormous expense of retrofitting what's already in place — a cost the industry would pass directly on to users — many experts believe that mitigation efforts should focus on new facilities. This is not to imply that a community that wanted to tackle this should not; taxpayers will vote for additional taxes or bond issues to underwrite things they feel strongly about, as will individual consumers.

There is also the issue of the loss of technological advancement if we were to limit new technologies to address their potential risks. What advantages should society sacrifice in order to limit EMFs? And what would be the economic loss of that decision?

EMF LITIGATION

Not unlike what has occurred in other environmental scenarios where a public health concern has collided with powerful economic interests, the EMF battle is being fought from the bottom up — community by community, or, where attorneys are involved, case by case. In the words of Michael Withey, a Seattle attorney who has been involved in a number of EMF lawsuits, "We think the juries, the mothers of America, aren't going to allow this to continue." So far, the legal fight has been slow and arduous — and expensive. Recently a number of EMF lawsuits were withdrawn due to costs.

The first EMF case to hit the courts was Houston Lighting and Power Company v. Klein Independent School District, filed in the eighties in Texas. The suit involved a dispute over the proposed siting of a 345-kV transmission line next to three schools. In the course of the litigation, HL&P built the line. The judge ruled they had to take it down, and the jury awarded the district punitive damages of $350,000. In another HL&P case, Bicki v. Houston Industries, the court also ruled against the utility — which had built a transmission line near a school, exposing students to 10-mG fields — and accepted expert testimony linking EMF exposures to cancer.

The first EMF cancer lawsuit where the plaintiff prevailed was John Altoonian v. Atlantic Electric, settled in September of '96. A jury had found the utility liable for causing the couple emotional distress — but not for John's leukemia — from a power line on their property and ordered it to pay $950,000. The utility appealed, then settled out of court — with no finding of EMF health risks. (By settling out of court, the industry can avoid establishing any legal precedents and also can keep a gag order on the case.)

A 1993 case, Criscuola v. Power Authority of the State of New York, advanced the claim that public perception of a hazard — in this case, EMF

— constitutes grounds for property devaluation litigation. The court found the property owners were entitled to compensation due to power line health fears, even without any conclusive scientific proof.

Most of the EMF litigation in the courts right now involves siting disputes or property values. There are many inverse condemnation cases that claim property has been devalued by proximity to power lines. Sage Associates, a California environmental consulting firm, has provided litigation support for a variety of these lawsuits, including SAN DIEGO GAS AND ELECTRIC v. DALEY, which established the "fear of EMF" as part of California case law.

Daley was a typical condemnation suit, except that the complaint stemmed from proximity to EMFs from a transmission line. SDG&E had taken a 200-foot easement for a 500-kV line from the Daley Corporation's Rancho Jamul property. This process is called EMINENT DOMAIN; the legal assumption is that a person will be justly compensated for this sort of taking by an arm of government. The question of just compensation, however, is often the subject of dispute. A further issue in this case was whether the property owner was entitled to severance damages because the easement had destroyed the best use of the rest of his property.

The utility offered Daley $110,000 for land Daley said was worth $1,450,000. Daley also claimed the prime property had been devalued by the power line and could no longer be developed as planned. SDG&E took Daley to court. A jury awarded Daley $190,000 for the easement, plus $1,035,000 for damages. On appeal, the award was reduced to $365,000.

According to Sage, the Daley case established "fear of EMF" as a compensable damage in condemnation actions. "From that time forward, evidence of electromagnetic impact was determined to be properly admitted in a condemnation suit." Writing in the *Real Estate Law Journal* in 1991, Sage established "fear of EMF" as case law in California.

Sage explained, "It's no longer possible for any utility to claim they don't know EMFs are dangerous. They need to prove they've made disclosure to people who are exposed to their lines. If they've written an Environmental Impact Report that says there's a significant danger to health, they can't continue to act as though there's no potential for significant impact. But the utilities are still denying this has a potential for damages. I tell them to go out and relate to the public concern. They're going to find that the marketplace isn't standing still when it comes to the perceived dangers of EMFs."

Two Florida inverse condemnation lawsuits — FLORIDA POWER AND LIGHT v. JENNINGS and FLORIDA POWER AND LIGHT v. ROBERTS — claimed that public fear of EMFs had an impact on property values and won. In 1987, the Supreme Court found that "impact of public fear (of EMF) on market value of property was admissible without independent proof of reasonableness of fear; the fear at issue here concerns the public's perception of health and safety hazards arising from human proximity to 500-kV transmission lines..."

Many realtors today will agree that proximity to power lines and substations effects home sales and property values. Some of them even go out with gaussmeters to measure EMFs and assure clients that the homes they're purchasing are safe. One agent in Texas estimated that houses near transmission lines have dropped more than 25% in their market value in two years.

In 1995, EMF litigation experienced a severe setback with Covalt v. San Diego Gas and Electric, a property devaluation case. The judge dismissed the case, ruling that EMF issues are under the jurisdiction of the PUC, not the courts. He also ruled there was no scientific proof that EMFs cause cancer. With this ruling, EMF litigation became much more difficult in the state of California.

In the opinion of Arthur Bryant, of Trial Lawyers for Public Justice in Washington, D.C., "Litigation in this area is by no means over. The industry is trying to make it look like there's no scientific proof (of EMFs causing cancer) — but the fact is, no one has proven yet that EMFs are safe. It's just going to require that attorneys be more cautious and careful in selection of these EMF cases. Our office is still pursuing two EMF cases, Bullock and Walston."

John Ward, of the Baltimore, Maryland, law firm of Quinn, Ward, and Kershaw, has filed two personal injury EMF cases against Northeast Utilities. His clients are two young people with brain tumors — Melissa Bullock and Jack Walston — who lived on the same block in Guilford, Connecticut, adjacent to an electrical substation. Ward agrees that the first few cases like this are going to be fought hard by the utilities. But he is also optimistic:

> I think they're going to be drawn out, with paper upon paper and the utilities filing delay after delay. But I'm very optimistic about winning the case. These tumors are closely keyed to EMF exposures. Neither of our clients had any other sort of exposures. We're claiming negligence in the siting of the transformers and in the utility's failure to monitor for EMFs or warn people of the possible dangers. This is a violation of state statutes that require a high standard of care for a utility.
>
> And the times have changed where the utilities can claim there's no scientific evidence. I recently gave a talk before a group of utility people. I pointed out to them that if they did their own reading of the studies, with an open mind, they'd see that forty-

two or forty-five out of fifty-two studies reported accelerated cancers with exposures to EMFs.

The evidence has gotten too strong to ignore it. I warned them not to make the mistake other industries have made — don't simply stand up and deny it.

Ward disagrees with the oft-heard argument that litigation itself may be standing in the way of voluntary mitigation on the part of the utilities, who perceive that either a more stringent standard of safety or their mitigation activities would constitute an admission of liability on their part, ushering in a landslide of damage claims. He says, "When a utility is sued in a tort you either sue on the basis of strict product liability or on negligence. You show that they had the duty to take reasonable steps to protect and they failed to take steps. If the utility can show they did this and that, I think it helps their case."

Peter A. Bell, professor of law at Syracuse University Law College, and others believe that, in the end, it will be the regulatory process, not the law, that will force changes on the utilities:

If the data are enough for you to start winning lawsuits, it's enough to scare the heck out of everyone. Then it's the regulatory process, not tort law, that will make a difference. People are going to get scared enough and angry enough to be all over the regulators. What I'm saying is, in tort law, if the science is strong enough to get over the causal incidence, then, as a society, we know that so many brain cancers, say, are going to be caused by EMF exposures. People are going to start screaming — and the regulators will respond. In that world, you don't need tort law to force the regulators to protect the public.

Professor Bell is probably right. In the long run, the voices of everyday people are what will make a difference in forcing regulators and decision makers to take the issue of the health hazards of EMFs seriously enough to take action to limit the public's involuntary exposure to EMFs. There are also a number of interesting cell phone lawsuits. (See Chapter 8.)

Chapter 6

EMFs at Home

> *Yes, I am concerned. When I do measurements in a home and I have a parent with two kids, and I can't give them an easy answer, I feel a wrench.*
>
> STUART MAURER, PH.D., New York Institute of Technology and ESEERCO EMF research program

New York Attorney Frank Ewing said he first became aware that there was "some sort of electrical problem" in his expensive co-op on Manhattan's Upper West Side when he and his wife, Brigetta, started hearing "crackling sounds in the air around us." Then his wife began to feel ill when she spent a lot of time in the apartment. Frank also had days when he suffered at home with headaches and a general feeling of malaise. "I'd be okay when I went out to work," he recalls, "but whenever I had a day off and I stayed home, I started feeling very strange. We were getting headaches and leg cramps; we never could sleep right; in general, we were feeling pretty lousy."

Ewing had seen a TV program about electromagnetic fields, so he decided to have the fields in his apartment measured. He was in for quite a shock: "The EPA found levels in my apartment of 6 to 22 mG. It turned out the building transformer was right next door to us, in the next apartment, and it was putting out these incredibly high fields. The Bureau of Electrical Control came out and cited the landlord, but he didn't fix the transformer."

In his quest for information and for some help with the problem, Ewing says he personally contacted "every agency I could think of that would have some answers." None of them was willing to come forth with suggestions. Consolidated Edison told him there was no problem. Says Ewing, "I just

WARNING: the electricity around you may be hazardous to your health

can't believe the health department or some regulatory commissions won't say, 'There's a danger. Get out of there.'"

When they couldn't get advice in the United States, Brigetta called her ex-employer, a big hospital in Sweden, and told them the story. Her friends in Sweden said they'd look into it for her. They called back a few days later and told her, "We've checked into this. Those fields are very high. You better move."

After failing to get their landlord to do anything to correct the problem, the Ewings withheld their rent at the advice of counsel and demanded that the transformer be fixed. Instead, the landlord took them to court and evicted them. When Ewing went back to the court to try to recover the cost of the move — "The entire thing has been an absolute disaster in our lives. It cost us a great deal of money and took a real toll on us emotionally" — the judge told him he didn't know anything about EMFs and couldn't waste the court's time with a case like this.

In New York, the owner of very large rental properties like the one the Ewings were living in is responsible for the wiring and the transformers on the customer side of the service drop. In other words, the power company brings electricity up to the building. After that, it's up to the owners to distribute the power to their tenants. The work has to be done according to the city code, but as in this case, enforcement can be lax.

Ewing had his new apartment measured before signing a lease. But the old apartment building is still on his mind. For all he knows, that transformer is still emitting high magnetic fields and people living in the building are in danger. He said, "A lot of people are facing great uncertainty on this. We don't want to be guinea pigs. What about the tenants in the building now? They can't all afford to move. What about public protection?" Frank Ewing now spends much of his free time attending public meetings to try to educate people about the hazards of magnetic fields: "It looks like, by virtue of what's occurred, we may spend years informing people about this problem."

Like the Ewings, a lot of us have decided to take matters into our own hands by limiting our exposure to magnetic fields as much as possible. People all over the country are calling in testers or purchasing gaussmeters and measuring magnetic fields themselves. By measuring the fields in your home you may derive peace of mind from learning that you're not being unduly exposed. According to EPRI, fields in residences generally range from 0.5 to 10 mG. But a number of recent studies have found that most residences have readings below 0.5 mG. A large-scale study in Denver showed magnetic fields to be in the 1.82-mG and higher range for only the top ten percent.

EMFs at Home

Measuring the fields in your home won't give you the whole picture, of course, because most people experience exposures from many sources. For adults, exposure comes from a combination of fields in the home, workplace, and transportation site.

HOW TO MEASURE/WHERE TO MEASURE

The California Department of Health Services has put together an excellent "Suggested Protocol for Measuring 60-Hz Magnetic Fields in Residences" for homeowners who want to do the measuring themselves. The following guidelines are extrapolated from this informative document (which is available from the California Department of Health, 2151 Berkeley Way, Berkeley, CA 94704). The protocol suggests you use a data sheet and a sketch pad to record readings as you go along (see Figure 6-A). Write your name and address, date, and time of measurements at the top of the data sheet. Then make a rough sketch of the parameters of your house and property in the space provided. Include the location of any nearby electric company sources like distribution lines, transmission lines, or substations near the property and house. This will be useful later on when you try to determine the actual source of any high fields you may locate.

Now you're ready to measure. (Instructions on using a gaussmeter can be found in Chapter 5.) The protocol directs you to begin outdoors, in your yard:

- Record readings at the four opposite corners of the house, 3 feet from the ground and the wall of the building. Walk along lines perpendicular to the walls of the building. Begin at the wall and stop every 6 feet or so to record readings on the sketch.
- Near the electrical service, walk a line roughly perpendicular to the service drop entering the home. (This is where the outside power lines are attached to your inside wiring; it's also where your main service breaker and electric meters are located). Recording readings about every 6 feet.
- This same technique can be applied to obtain profiles for distribution lines on or near your property.

Conducting a Room-by-Room Field Survey Now you're ready to measure the fields inside your house. You may want to take a couple of readings, then average them. Certainly, if you come up with any dangerous field levels in your home, you will want to check them again by repeating the measurement.

WARNING: the electricity around you may be hazardous to your health

RESIDENT		TECHNICIAN	
Name:		Name:	
Address:		Address:	
City, CA Zip		City, CA Zip	
Phone:		Phone:	
Date:		Company name:	
Time:		Meter type:	

FREEHAND PLOT PLAN — YARD, HOUSE AND POWER FACILITIES

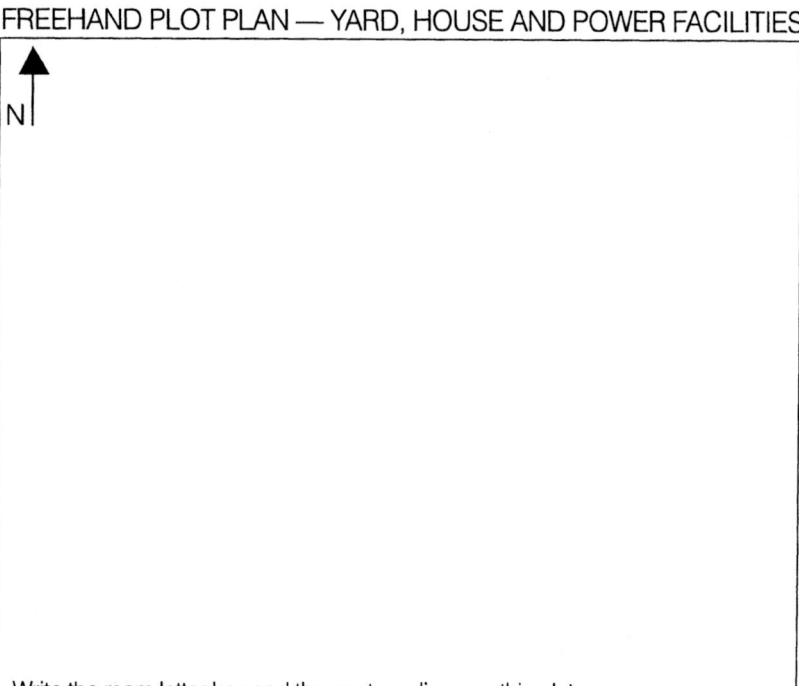

N ↑

Write the room letter key and the spot readings on this plot.

(Key Room)	Center
(A) Front Door	
(B) Family Room	
(C) Kitchen	
(D) Bedroom 1	
(E) Bedroom 2	
(F) Bedroom 3	
(G) Bedroom 4	
(H)	
AVERAGE (Mg)	

Appliance (On)	Room	4 in.	18 in.

Figure 6-A. Data Sheet. (*Source*: California Department of Health)

EMFs at Home

Always measure during times of peak usage. Evening, around dinnertime, when most people are at home using electricity, is a good time to conduct your survey. (As you have seen, if you live in a city, the electricity in use in other homes in your neighborhood can have a major effect on the fields in and around your house.)

Start your measurements at the front door. Be sure to measure the service drop — that is, the point where electricity from the power system is hooked up to your wall wiring (the box or panel where your main switch and fuse box and meters are located). Be sure to measure on the other side of the service drop if it's on an inside wall. You may find high fields there because ordinary walls won't shield a magnetic field. If you discover a piece of furniture — maybe a chair — located n the middle of those relatively high fields, it would be a good idea to move it elsewhere. If the porch is on the other side of the service entrance, be sure to measure there as well. (See Figure 6-B)

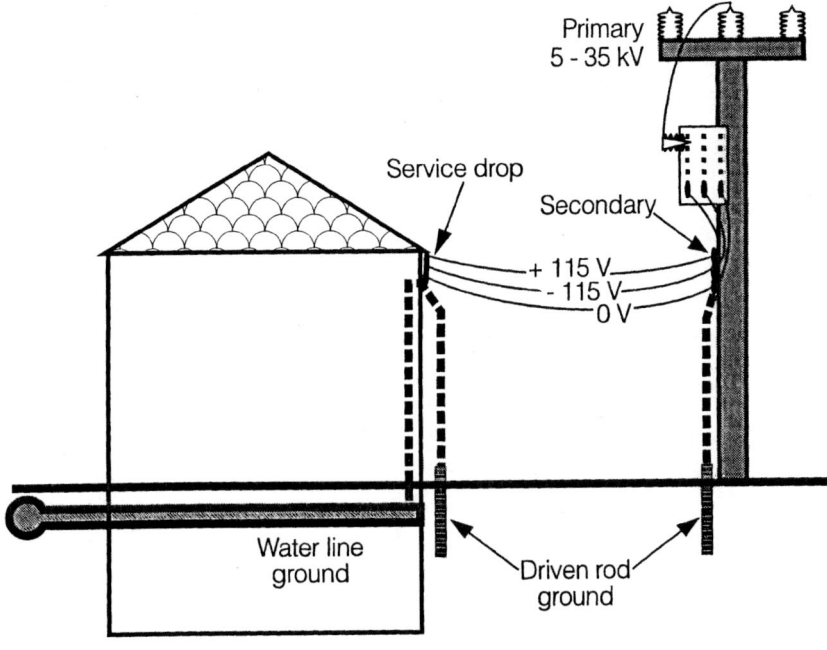

Figure 6-B. Typical home service drop. (*Source*: **U.S. Office of Technology Assessment**)

WARNING: the electricity around you may be hazardous to your health

Next, move on to the most frequently used rooms in the house — bedrooms, kitchen, family rooms. Take measurements in the center of rooms, about 3 feet from the ground. In bedrooms, measure eight inches above the center of each bed. Be sure to record the level of fields as you go along. It may help to have someone else do this with you. Later on in the chapter, we'll talk about identifying the actual source of any high magnetic fields you detect.

As you conduct your room-by-room magnetic field survey, make it a point to always measure the most frequently used areas in each room, as well as the center spot. Pay special attention to favorite areas of a room. If you do discover high fields around them, change your favorite spots. Pay special attention to where your children spend most of their time and check those fields carefully — they may be playing or sleeping in high fields. If this is true, rearrange the rooms to get the kids out of those fields.

While you're measuring the magnetic fields in the various rooms of your house, you should also be doing an inventory of the electric appliances in each room. (See Appendix C for a list of magnetic fields from common household appliances.)

It has been determined, in a number of studies, that the overall magnetic field level in a home is not greatly affected by home appliances. But exposure may very greatly according to personal appliance use. All home appliances that plug into the wall create magnetic fields. (Battery-operated appliances don't have fields, but their rechargers do.) Many home appliances create very high fields — for example, electric blankets (50-100 mG or 5-10 mG for the new "reduced field" models), hair dryers (3-1,400 mG), electric shavers (14-1,600 mG), toasters (10-60 mG). Some particular appliances — electric blankets, waterbed heaters, black-and-white televisions, hair dryers — have been linked to cancer and reproductive problems in a number of studies.

Appliances emit magnetic fields all around them, not just in front of them. In particular, many large appliances have very high fields in back, where the motor is located. And you can't judge the field by the size of the appliance. You have to measure it. Many small appliances have surprisingly high fields because of the types of motors inside them. Be sure to measure all around the appliance to discover where the motor is and where the highest fields are being emitted.

The important thing to remember about appliance fields is that the farther away you get from the source, the lower they become, until they disappear. In the case of the high magnetic fields from appliances, the good news is that they drop off dramatically as you move away from the source. And you don't have to be very far away to escape their fields — generally just a foot or two will do. According to the research, ninety-five percent of measured appliance fields one foot away from the source were only 1 mG. Five feet away, the fields measured even less than 1 mG.

Three is the magic number when we talk about safety from appliance magnetic fields. The rule: Keep your distance. Unlike other kinds of magnetic fields, fields from appliances decrease greatly as you move away from them — even only a few feet away. Three feet is a very good safety zone to establish between yourself and the average electric appliance. Unfortunately, this isn't always possible because many hand-held appliances have to be operated at close range.

There are also great differences in the duration of exposures from various appliances. Many small, hand-held appliances are only used for a matter of minutes. On the other hand, people sit in front of the television for hours every day.

Unfortunately, except in the case of a handful of electric blanket manufacturers, there are currently no low magnetic field appliances on the market. Electric appliance manufacturers, as a group, have taken the position that there is no proven danger from their products, so they aren't taking the initiative to reduce their products' fields. When manufacturers say changes would be too expensive and would be passed on to the consumer, remember that the American automobile industry completely retools its factories every two years to change the design of its cars, as well as to increase safety, and people still buy cars. According to Dr. Wendell Winters of the University of Texas, reducing appliance magnetic fields wouldn't be that costly: "It's up to the people responsible to redesign these things. In many cases, they've found that this can be easily remedied and without great expense. Magnetic fields can actually be eliminated from appliances rather easily, just by changing the way they do the wiring." Another recourse would be to replace the cheap motors in many small appliances.

Mary Gillespie, assistant director of communications for AHAM (Association of Home Appliance Manufacturers), says the reason manufacturers aren't redesigning their products to lower fields is "because there's no scientific evidence linking this lowered magnetic field to a healthier life." A 1989 printed industry response on the subject of "Home Appliances and Electric/magnetic Fields" is still being sent out by AHAM to customers and the media. The paper states, in part: "To our knowledge, no health effects have been verified. The scientific community's current consensus is that no definitive evidence of a cause and effect relationship exists" and "appliance companies are monitoring the situation regarding health hazards. As further health data and analyses are generated, our industry will respond appropriately."

WARNING: the electricity around you may be hazardous to your health

Until that time it certainly makes sense for you to make your own decisions about the individual appliances that are indispensable in your own home. For example, Dr. David Carpenter says, "Today, no responsible person would have an electric blanket or use a hair dryer on a child."

Let's move to the bedrooms. Begin by measuring eight inches over the beds. If there are unsafe levels where you sleep, start moving the furniture immediately. Find an area with a lower reading and put the beds there. Also look out for two of the most dangerous home appliances that are generally found in the bedroom — electric blankets and waterbed heaters. A number of studies have linked both of these to higher rates of miscarriages, brain tumors, and leukemia. Wertheimer and Leeper, who measured 15-mG fields from electric blankets and 3- to 4-mG fields from waterbeds, reported that children whose mothers used electric blankets during their pregnancies were two and a half times as likely to develop brain tumors, and when the mothers used heated beds, they had a higher rate of miscarriage. The hazard is compounded by the fact that people generally spend seven or eight hours a night lying in the field. A further problem comes from the fact that the thermostats on these items keep switching on and off all night long, exposing the user to magnetic field surges. The only way to eliminate electric blanket fields is to unplug the appliance before you get into bed. (Electric heating pads pose the same problem.) Waterbeds are even worse because their heaters must operate yearlong. Dr. Fred Stertzer, of the New Jersey Commission of Radiation Protection, says,

> It's very easy to manufacture an electric blanket where the magnetic field is reduced by 95 percent. It's an almost trivial change and doesn't cost any more. It consists mainly of running the wires in opposite directions so that the fields cancel out. All the new blankets have these reduced fields. The problem is, huge numbers of people are still using the older models. I'm for writing strict regulations on these things, warning people not to use them. You shouldn't let people sleep under them. They alter the nighttime melatonin production (see Chapter 4). In general, you should avoid anything that has a profound influence on a major hormone. It really behooves public officials to do something about this.

If you have an older-model electric blanket, stop using it, or use it, but don't plug it in.

Another source of high exposure in the bedroom is the old-fashioned analog clock — the kind with hands. (Digital clocks and battery-operated models don't emit magnetic fields.) Again, the problem is one of high field plus duration of exposure. You sleep all night long with the clock right next to your head. The solution is simple: Move it at least three feet away, or replace it with a battery-operated clock.

Radios, stereos, and televisions (especially if you still have an old black-and-white model) should all be measured. The same goes for a portable phone, if you happen to have one on your bedside table.

Don't forget to check out the fields produced by electric heat sources. You may have electric baseboard heating or use one of those portable electric coil heaters. Measure the fields around them, and if they're high, situate them so that you and your family aren't sleeping in those high fields. It's not unusual for parents to put their baby's crib or the children's beds close to the heat to keep them warm. It's been reported that a baby with leukemia in Seattle had spent a year sleeping in a 6-mG field near a portable electric heater. To determine where a field ends, just step back slowly and keep measuring until your meter settles at a safe level. Stand still for a moment to make sure you have the true measurement.

A number of studies have linked cancer and miscarriages to concrete-slab coil heaters — the kind that are built right into the ceiling or floor and regulated with a thermostat. If you have this kind of heating in your home, measure above it if it's in the floor and below it if it's in the ceiling. And remember to measure in the rooms above, below, or adjacent to electric heaters, because magnetic fields can go right through walls. That holds true for large appliances, also. If you have a television positioned against a wall, for instance, measure the field that may be generated in the next room, from the back of the set.

The bathroom often has a number of small appliances with very high magnetic fields, including some models of electric toothbrushes (are these really necessary?), electric shavers, and hair dryers. The bathroom lights, especially if they're fluorescent, are another potential source of high fields. In this little room, it's the cumulative effect you're after, so measure with everything turned on. Then measure around each individual appliance. In the case of battery-operated appliances, remember to measure their rechargers.

Hair dryers are a known source of extremely high fields; they are one of the appliances named as hazardous in some of the research. Whether or not you continue to use your hair dryer is a personal decision. If you do, the farther away from your head you hold it, the more you'll reduce the field — which drops away greatly with distance. (See Appendix B for some figures on this.) The good news is that companies are beginning to market safe hair dryers similar to the ones you find mounted on bathroom walls in

WARNING: the electricity around you may be hazardous to your health

hotels. The wall mounting feature allows you to use the dryer and stay completely out of the field, since it's the motor that puts out the EMFs. Bloomingdales offered one such model, the Andis Pro, in its 1997 winter catalogue.

The kitchen is a very likely spot for high magnetic fields. (Be sure to take your reading while dinner is being prepared.) The cause of the high fields is the large number of electric appliances generally going at the same time — all those labor-saving devices that have been so successfully marketed over the past three decades. Measure around some of those appliances and you may be in for a shock — not literally, of course. In particular, turn on all the burners and the oven of the electric stove and measure the fields. I have measured fields from 20 to 100 mG around electric stoves with only two of the four burners turned on. The electric stove in my apartment right now generates fields as high as 30 mG when it's turned off! I attribute this to a wiring problem or some sort of faulty circuitry, although an 11-mG field is coming from the analog clock/timer on the front of the stove. (A rule of thumb is: The smaller and cheaper the motor, the higher the field.)

Then consider the cumulative effect of the magnetic fields from the refrigerator, lights, microwave (not the microwaves but the EMFs, coming from the transformer it runs on), and all the other kitchen appliances that may be running during the preparation of a meal. Also remember that children love to hang around close to their parents while dinner is being made.

So you can sense the dimensions of what Lynne Gillette, of the Environmental Protection Agency EMF group, calls a "sex-bias magnetic field problem," since women often spend more time in the kitchen than men. While Gillette was conducting her survey of the daily magnetic field exposure of people in her office, she ran into this controversial finding: "I found what seemed to be some sexual bias about magnetic field exposure. Whatever people tell you, it turns out it's still women who are the ones preparing dinner, and for the hour or so they're in the kitchen, they're next to a variety of kitchen appliances with very high fields."

A rule to follow with regard to the magnetic fields from appliances is *keep your distance.* Break your habit if standing over the toaster waiting for the toast to pop or standing next to the coffee maker as it drips away. In fact, stay a yardstick away from an appliance whenever it's possible to operate it at a distance. Remember three feet is generally the magic number. (Of course, you don't need to carry a yardstick around. Just measure your arm when it's fully extended and get a sense of how close that is to three feet. You can teach your children to "walk off" the distance the way workmen do.)

Unfortunately, it's impossible to keep your distance in the kitchen where so many appliances have to be operated at close range where you're right in the magnetic field. On the one hand, there are appliances you can turn on and walk away from. And even if you can't do this with your electric stove, you can disconnect its analog clock. Do this and then measure the fields. If they're still high, you're going to be faced with a potentially expensive personal decision: whether or not to trade in the offending electric stove for a gas range. While you're debating it, consider a woman leaning over the stove while she's using all four burners. We should make a note that the number one cancer death for women today is from breast cancer. And, according to researcher Richard G. Stevens of Pacific Northwest Laboratory, breast cancer and breast cancer mortality are on the increase right now. The connection between time spent in kitchen magnetic fields and female cancer trends certainly bears looking into.

An obvious next question is: Do you really need all these labor-saving devices if they're a potential health hazard? In that context, I can think of many I can do without and probably you can also.

Microwave ovens have become an essential item in so many homes. There are two sources of radiation to worry about from these ovens. The most obvious is the microwaves themselves, which are high-frequency radio waves. This was a major concern with the early ovens, which were found to leak microwaves so badly that *Consumer Reports* recommended against them in 1973. Presumably, the situation has been improved by the manufacturers, but certain models can develop leaks, and you are advised to have your oven checked every year by a licensed serviceperson. Microwave ovens also emit 60-Hz radiation like any other home appliances, from their small transformer-motors located in the back, so be sure to measure around yours for the 60-Hz magnetic fields.

Keep children away from the microwave. Unfortunately, many children are growing up with the notion that the microwave oven is a toy. There are horror stories of children playing by testing out various items in the microwave and practically leaning their noses on the door of the oven, to see them cook! The oven door is the most dangerous location for microwave leakage, but 60-Hz fields can occur all around them, especially in the back. Right now there is a very unfortunate trend of marketing food items designed especially for kids to cook in the microwave. If you must have a microwave, you and the children should be very careful to stand clear of it — at least five feet away — while it is on. I suggest you turn it on, then cross the kitchen until it goes off.

In the living room, the home entertainment center can be the site of some high appliance-related magnetic fields. Measure around the television, the VCR, the stereo or CD player, your personal computers (especially their

WARNING: the electricity around you may be hazardous to your health

video displays), and whatever else your family has invested in to keep everyone entertained. By the way, television interference or unexplained, poor reception may be an indication of a high magnetic field in a room. These fields vary greatly. Measure them all, then act accordingly. Don't forget the sources of lighting in the room, the lamp cords, switches, and outlets.

One hopes you won't find any unduly high fields around the electronic components in your living room. But if you do, again you have two choices. One is to establish some clear physical boundaries to protect your family from exposure. Lots of people these days have run a line of tape across the floor 3 feet away from the television, to keep the children back.

The other solution is to try to reach some sort of agreement about what constitutes a necessary electric gadget. Simply put, what can you do without? For example, portable radio phones produce inordinately high magnetic fields. Worse yet, you hold them against your head all the time you're using them. Recall the studies about the pineal production of melatonin and how dangerous the effect of magnetic fields on this hormone is believed to be. The pineal gland is in the middle of your head, right where the ancients in their abiding wisdom located the mythical third eye.

The same holds true for cellular telephones. Not only are they exposing the people who use them to magnetic fields applied directly to their heads, but other people are being involuntarily exposed to their transmission relay magnetic fields. (See Chapter 8.)

Now you've completed a magnetic field survey inside your home. You've measured at the service entrance to the house, in the center of the family rooms, near the beds, and around the most frequently used appliances. What if you actually do identify some high magnetic fields, or hot spots, as they're often called? Your next step will be to do some careful detective work to try to ferret out their sources. This can sometimes be a very difficult task, so you may decide to call in a professional to assist you.

HOW TO FIND THE SOURCES OF HIGH FIELDS

The first thing you need to know is whether the high fields are coming from sources inside or outside your home. You can pretty much eliminate sources inside your house by simply throwing the master electric switch to shut down all the circuits in the house, then repeating the reading. If it is still just as high, then the problem is coming from an outside source.

Luciano Zafanella is the manager of EPRI's High-voltage Transmission Research Facility in Lenox, Massachusetts, a model neighborhood that has been built by the electric industry to analyze the ambient electric power sources of magnetic fields and develop engineering techniques for field management of those sources. Dr. Zafanella explained, "The two most common sources of magnetic fields in this country today are the power lines outside on the street and the home grounding system." A pilot study conducted by EPRI in 1987 listed the primary sources of residential magnetic fields as:

- Transmission lines
- Distribution lines
- Currents in the residence's grounding system
- Unusual wiring arrangements in the residence
- Appliances (Appliances don't contribute greatly to the overall high magnetic fields in a house.)

Take a look at the power lines around your house. We've already considered transmission-line magnetic fields at some length in Chapter 6, but certainly if your home is located near a transmission line right-of-way and has high fields, you should find out whether those fields are being caused by the transmission line. To do that, call the utility company and have them come out and check for you or hire a private electrical contractor.

If there aren't any high-voltage transmission lines near you, and it's likely that there aren't, be sure to take a look at the distribution lines and pole-mounted step-down transformers around your home as possible sources of those high fields.

To survey the lines yourself, take readings at the side of your house that's nearest the line and compare them with readings taken on the opposite side of the house. Use the four-corner measurement as a standard. If you note a difference in the corners near the line, it's a good guess that the fields are coming from there. Also measure as you walk from the lines to your house to try to get some idea of the fields.

But remember, distribution-line fields are extremely complicated, and you're likely to need outside help to trace these fields. Many lines, in particular the high-current feeder lines from substations, are buried and hard to locate. If the power lines are emitting high magnetic fields, this is the result of imbalances caused either by the load current on the line or by the total of the net currents. (Appliance loads in your home create both negative and positive currents, in unequal amounts. The difference between these is net current, which has to find its way out of your house, to the power company transformer. As we shall see, the current sometimes goes out on the neutral line from your service drop, but more often it takes a variety of other routes that may create magnetic field problems.) A professional can detect the dif-

WARNING: the electricity around you may be hazardous to your health

ference between these, using the rule that load current magnetic fields decay within a predictable distance from the lines, but net currents disappear much less rapidly. (See Figure 2-A for a typical distribution service drop.)

In any case, if you suspect that the source of high fields in your home is a distribution line, it's a good idea to involve the power company in your concern. Take the hypothetical case of a person who measures 10-mG fields in his home. I asked Greg Rauch, who used to be the project director for a number of EPRI magnetic field management programs and is currently a consultant with Electrical Research and Management, what he would do if he had discovered high fields like these in his home. He said, "If a person found a magnetic field that measured 10 mG, say, two or three meters away from a power line, you'd be justified in having the power company come out and check the neutral from the transformer. A high reading like this is a situation that needs to be remedied. If you're getting a return of 10 mG, you have a problem."

So, you're perfectly justified in asking your utility to come out and check its lines near your house as possible sources of high magnetic fields. And the utility should be able to find out what the fields are and determine their source. For one thing, the company can monitor whether or not the lines are balanced. Ideally, the distribution lines are all supposed to be balanced, but this is actually very difficult to accomplish. The utility company should also check for broken connectors and other physical causes for those high fields. Depending on where you live, the utility may or may not be willing to come out and measure the fields and help you determine their source.

A company such as San Diego Power and Light, which has been slapped with two significant EMF lawsuits, is receptive to requests from its customers. According to SDP&L public relations director John Britten, SDP&L has set up a special department to help customers identify high magnetic fields. The company recently purchased a hundred magnetic meters and will send employees out to conduct residential measurements and help customers locate the source of high fields. At one point in 1991, SDP&L printed an announcement about this service in its newsletter and received 987 requests for measurements. Britten says that, although it couldn't make any sort of recommendation based on its readings, the company was able to fill all those requests.

If you decide to assess the electric landscape around your home by yourself, you may want to try using the wiring code configuration protocol designed by Drs. Nancy Wertheimer and Ed Leeper and used in a number of epidemiological childhood cancer studies to estimate chronic residential exposure. Since the protocol wasn't originally designed to be used by laypeople, it's included here mainly as a guide for visually identifying potential high-field sources. It provides an easy method for checking out the overhead power lines and transformers near your house as potential sources of high magnetic fields. Remember that *primaries* are the wires that are at the top of the poles.

If you determine that your house falls into one of the high-current configuration (HCC) categories, and you measure high fields inside the house, you can be more aggressive about having the power company come out to check their equipment for possible sources of those fields.

DO YOU LIVE IN A HIGH-CURRENT CONFIGURATION HOUSE?

According to many studies, children are more at risk for cancer if they live in homes with the following electrical wire configurations:
- The first or second home away from a pole-mounted step-down electrical transformer.
- Homes located less than 40 meters from large-gauge primary distribution wires. These are the thick wires found at the top of wooden poles in neighborhoods. They are connected by porcelain insulators.
- Homes less than 40 meters from an array of six or more thin primaries (also on top of the pole).
- Homes less than 20 meters from arrays of three to five thin primaries or high-tension (50-230 kV) wires.
- Homes less than 15 meters from secondary wires (240 V) that run *directly* from a pole-mounted transformer without losing any power in a service drop (a connection to a home) beyond the pole before passing your house.

As we saw in Chapter 5, the utilities do have methods at their disposal to lower the magnetic fields around their power lines. But, in the case of unbalanced currents on distribution lines (which may be the source of high fields in your home), this is difficult to accomplish. So, for the present, instead of waiting for the power company to eliminate the problem, you have another option: Make some immediate changes inside your home that will help reduce your family's exposure to the high fields coming from those distribution lines. In general, all this involves is doing some rearranging in the rooms with the high fields. For one thing, you can move the couch or beds to a position where the fields are lower, or you can even switch rooms. Of course, this will only work if the high fields are limited to particular rooms or parts of particular rooms.

WARNING: the electricity around you may be hazardous to your health

Another possibility is that the high magnetic fields in your home aren't coming from the power lines. In that case, they're likely to be coming from your home's grounding system.

Most homes in cities in the United States have a multiground system, which means the electricity is grounded to a ground rod of some sort that returns it directly to the earth and also to the water pipes. In fact, the National Electric Safety Code mandates this system to protect people from both electrical shock and fire that can be caused by arcing faults of sparks from the electric system. In other words, excess electricity is directed into the earth. The same is true of neutral lines from large appliances, like your washer and dryer. They have to be grounded to keep you from getting shocked when you come in contact with them.

The electricity you use to light your house and run your appliances comes into your house at the service drop. At the service drop, electricity runs on two hot or energized wires, and a third neutral wire is connected back to the power line. The neutral wire provides a path for extra electrical charge to return to its source (the transformer on the pole outside your home) to protect fire and shock. Unfortunately, a lot of these neutral currents continue to flow along the water pipes instead of simply returning to the electric power system as they're theoretically supposed to. The problem is that this neutral current running along the water pipes (not the water itself) often creates high magnetic fields in homes along the water system.

This will only occur if you live in an urban area where you share a utility and a water system that uses metallic pipe. Plastic pipe will not conduct electricity. If you have metal water pipes, neutral current may be circulating from the service entrance at your house to the water pipes, over those water pipes to the water main, and even along the main to your neighbor's water pipes, creating dangerously high magnetic fields in homes along the way.

In fact, in a test conducted recently in California, currents of up to 10 amps were measured on typical residential water pipes. Using the formula: MF equals two times the amps divided by meters, we get 20 mG fields in homes one meter away from these pipes.

In the same way, currents coming from your neighbor's house may be flowing on your water pipes and causing high magnetic fields in your home. These currents may fluctuate greatly — as will the magnetic fields — every time your neighbors turn something on or off in their house. This may provide a clue when you're trying to find the source of your home's magnetic fields: Varying high fields are likely to be caused by currents on the grounding system.

It may be easy for you to locate these currents with your meter. In fact, if you have a ground-current problem, your meter will lead you over the exact route the water pipes take as they carry water from house to house. Magnetic fields resulting from your home grounding system can be eliminated easily without going to any great expense. But it is very important that whatever is done is completely in compliance with the National Electric Code, first of all to protect your family from shock or fire and second to avoid jeopardizing your fire insurance coverage.

An electrician can change the grounding system to stop your neutral current from contributing to the plumbing-line fields. Then, depending on the location of the service box with regard to the water pipe entrance, he may also be able to fix the neutral grounding conductor at your service entrance so that it won't be a problem inside your house.

Another solution is to get a plumber to install a dielectric union — two conductors that are separated by a nonconducting or dielectric insulating material — on the water pipe at the property line shut-off to stop currents from other people's homes. The dielectric union has to match the particular metal used in the plumbing line on either side, with some nonconducting material in between. Again, be careful not to do anything against code. You may want to check with the local plumbing inspector to make sure that what you want to do is legal in your community. Although the connector itself will only cost about $20, this will not necessarily be a cheap fix because the plumber will charge somewhere in the range of $200 to $300, since it will take a day to dig down to the frost line to reach the plumbing line, and then install the union. If you can dig your own hole, you can save some money. Also, the water company may have to be called to turn off the water. But, the good news is that in one day you may eliminate the highest magnetic fields in your home. People with ground-current problems have reported indoor fields as high as 20 mG.

Also, be sure to check the current on the buried cable TV lines, because sometimes home grounding wires are connected to these cables as well.

The next possible source of high residential magnetic fields is unusual *wall wiring*. A professional EMF tester or an electrical engineer would have no difficulty tracing the high magnetic fields in your home to wall wiring; it's a very simple call to make. Some clues that the magnetic fields are coming from the wall wiring in your house are obvious and your gaussmeter will point to them. For instance, if the fields seem to form a loop around one or two walls in the room, you probably have a wiring problem. Another way of reaching this conclusion is by the process of elimination: If the fields aren't coming off the electric power system, and they're not taking the path of your water lines, they're likely to originate in the wiring. Wall wiring fields, however, can be very high and can extend throughout the room.

In any case, if you suspect that an unusual wiring situation is responsi-

WARNING: the electricity around you may be hazardous to your health

ble for the high magnetic fields in your home, you should call an electrician to help you find the source of the problem and correct it. Often, a magnetic field specialist can refer you to an electrician. Karl Riley is one such specialist, and Riley says EMFs from wall wiring are always the result of wiring errors — which can be remedied by redoing the wiring to code.

UNUSUAL WIRING

- Knob and tube wiring.
- Electrical loop problems caused by the improper wiring of two- and three-way switches.
- Subpanels that are wired in such a way as to cause loops in the wiring.
- 240-V appliances fed through 120-V wires.
- Any number of out-and-out wiring violations that could result in hot spots.

Unusual wiring is not necessarily against code, but it's often the result of someone taking the easiest or cheapest route in the job of wiring a house, resulting in a rather costly shortcut you now have to remedy. For instance, take the case of a simple two-way switch — the kind that's used so you can turn a light on or off at both entrances of a room. Sometimes, just to save a little wire or a little time, an electrician will run the neutral wire on a different wall from the energized wire. By doing this, a current loop has been created, with a resultant high magnetic field. To remedy the situation, someone must rewire the switches, making sure to run the neutral wire wherever the hot wire goes.

Knob and tube wiring is an old-fashioned method of wiring that isn't used anymore, but is still found in many older homes and buildings, especially in the northeast. It's not illegal, but because in this method the hot and neutral lines were enclosed in porcelain tube insulators, then run separately at some distance from one another — sometimes even on separate walls — a house with knob and tube wiring generally has magnetic fields two and three times as high as those typically found in homes with more modern wiring. The problem, of course, is caused by the fact that with the hot and neutral wires separated, they aren't able to cancel each other out, and so they cause current loops with high magnetic fields. If you own an older home, you can tell if it has knob and tube wiring just by looking at the service drop. If you aren't sure, call an electrician to check it. Obviously, correcting this particular unusual wiring problem may be quite costly, as it entails rewiring the whole house.

Branch panels, sometimes called subpanels, are usually added when a building is expanded, in order to provide additional electricity. They are located apart from the main service box and are often improperly grounded, in violation of the electrical code. They need to be rewired to reduce fields.

Sometimes, instead of a single 240-V wire, two 120-V lines have been used. The remedy for this, of course, is to change it. But, in general, if you find that none of the preceding wiring problems are the cause of the hot spots in your house, you're probably going to discover the wiring has not been done to code. Then, unfortunately, your home should be rewired.

You may have some difficulty, depending on where you live, finding an electrician who will come to your home and understand the magnetic field problem, or finding an electrician who is willing to correct the problem. In this case, you should try contacting one of the companies that offers electromagnetic field inspections. In the near future, these companies will probably have someone on the staff to do the kind of electrical work you need.

In the absence of any of this kind of unusual wiring, and if the wiring in your home has been done to code (wiring codes require that the neutral and active wires be run closely together to cancel each other's fields), your home shouldn't have any magnetic field problems resulting from wall wiring.

There's still one other situation involving residential magnetic field problems that we haven't discussed. What if you discover that your home has dangerously high fields, you even locate the source of those fields, but there's just no way for you to change the situation? Then you're faced with a very difficult decision. Are you going to continue living in those hazardous magnetic fields, or are you going to move?

Clearly, this is a worst-case scenario. Most people simply can't afford to abandon a home, especially if they own it. On the other hand, people may have to do just that if they feel their family's health is being jeopardized. Then there's also the question of selling the residence to another family: Can you do that in good conscience without warning them of the problem?

Chapter 7

EMFs in the Workplace

> *The public perception is that anything over 3 mG is dangerous. What about a secretary sitting in a 300-mG field all day long?*
>
> RICHARD TELL, of Tell Associates

In the spring of 1991, The Boston Globe took a giant step to protect its workers from on-the-job exposure to EMFs from their video display terminals (VDTs). At the direction of Dr. Terrence O'Malley, the newspaper became the first major workplace in the country to issue a medical directive aimed at reducing employees' magnetic field exposure by creating a safe zone around the office computers. The directive instructed everyone to move their desks "to establish a minimum of 3 feet between each individual and the back and sides of the nearest VDT," explaining that this was being done in order to achieve less exposure to weak magnetic fields, in case there turned out to be harmful effects from the pulsed low-level magnetic fields from VDTs.

Dr. O'Malley calls the decision a "trade-off," between the "small but potentially demonstrable health effects of VDTs, and the potential for associated hysteria about these effects." He feels that the move shows the owner's concerns about the people who work at *The Globe*. "We don't know what the risks are and it would take a long time to prove that. Meanwhile, we could do something simple and inexpensive to protect our employees." In opting to protect employees from the possible hazards of exposure to magnetic field from their VDTs, *The Globe* presents a model for other large companies, at a time when more than forty million computers are in use in U.S. workplaces.

EMFs in the Workplace

It can be said that today's high-tech state-of-the-art electronic workplace is a sea of electromagnetic fields. It can probably also be said that, like most of the writers, editors, and office workers at *The Globe*, the vast majority of the nation's computer users have little understanding of either the short- or long-term health risks they may incur by spending their days exposed to the emissions from the VDTs. Just as we've seen in every other area of electromagnetic exposure, a great deal of contradictory information about the health dangers from our computer technology has been bandied about, with the result that many people feel unequal to the task of protecting themselves from electromagnetic fields they may encounter at their jobs. Unlike *The Boston Globe*, most employers are satisfied to just sit out the controversy, instead of taking some simple steps to reduce their employees' risks from computer EMFs.

People began to worry about emissions from their VDTs in the early eighties, when a number of stories surfaced about clusters of women workers in the United States and Canada with increased miscarriages and birth defects that were caused by their VDTs. Studies also reported other symptoms — headaches, rashes, sleeplessness, blurred vision, fatigue — some of which had, in the past, been associated with other aspects of computer use, such as repetitive motion and the physical position of the operator. And for nearly two decades, studies had linked VDT use to cataracts, blurred vision, and eye strain. But these findings were all more or less ignored, until a major study in 1988 convinced people that it was possible that radiation from their computers was dangerous to their health.

The study was conducted at Kaiser Permanente Hospital in Oakland, California. It has been designed to assess the effects on pregnancy outcomes of statewide malathion spraying that had been carried out to combat a Mediterranean fruit-fly infestation. But in the course of their investigation, the researchers found that women who spent twenty hours or more at computers during their first trimester of pregnancy experienced two and a half times the miscarriage rate of women who didn't work at computers. Although the Kaiser study had been competent and carefully designed, it was widely disputed by special interests. Then in 1990, the National Institute for Occupational Safety and Health (NIOSH) studied telephone company employees and reported that there were no miscarriage risks connected with VDT use. But the NIOSH study had some problems: (1) It wasn't originally designed to measure EMFs at all; (2) both controls and subjects experienced equal 60-Hz EMF exposures; and (3) the researchers had only measured very low-frequency emissions, not ELF emissions, which would include 60-Hz electromagnetic fields.

Nonetheless, the NIOSH story ran in almost every major newspaper, coast to coast, with the result that many people today are still confused about the potential EMF dangers of working at a computer. Now and then, in response to complaints, the government has occasionally measured VDT emissions and found high magnetic fields around particular computers, but no American government agency has addressed the problem as an occupational health issue. This year, the Institute of Electrical and Electronics Engineers (IEEE) announced a standard for VDT emissions, but like other so-called EMF standards, this one is only voluntary.

A few countries have tried to do more, however. In 1983, the Canadian Center for Occupational Health and Safety recommended that office space be arranged so that workers wouldn't be exposed to emissions from the sides and backs of other people's computers. Sweden, always in the forefront of magnetic field regulations, has established a manufacturing standard for both ELF and VLF emissions from VDTs. For ELF radiation, the limit is 2.5 mG, at twenty inches. Many computer manufacturers in this country are already following the Swedish example and lobbying for the government to set a similar level here.

Just as we've seen with the overall EMF health risk issue, the question of VDT emissions has been obfuscated for years by special interests, and in this country it has been subjected to a virtual news blackout. When clusters of workers with computer-related illness have been reported, they've been given short shrift by the press, or coverage has focused on official rebuttals of the accusations that emissions from VDTs were responsible. Critics complain about the seeming disinterest of the American press for the VDT health effects story. But Louis Slesin, editor of *Microwave News* and *VDT News*, feels it's more than that:

> I think the problem is there are great cross-ownership problems between newspapers and broadcast properties. For instance, NBC is owned by GE. And GE is up to its ears in production of everything from power-line equipment to home appliances. What's interesting in this is that in other cases the press acts as ombudsman — reporting on cover-ups. But here, they're part of the problem. When it comes to VDTs in the newsroom, or a question of the industry protecting broadcast signals from regulation, the usual advocacy isn't functioning. Even in the case of the unions, the Newspaper Guild doesn't want to upset the apple cart and cut down on jobs. It's such a major problem that no one wants to touch it.

EMFs in the Workplace

Despite the smoke screen, however, there has been some progress in dealing with EMFs in the workplace. For one thing, a few businesses, like *The Boston Globe* and the Fund for the City of New York, have restructured their offices to keep workers away from computer emissions. Thanks to an agreement that New York City signed with the American Federation of State, County, and Municipal Employees, their government workers are required to sit at least forty inches from each other's VDTs. Some companies have even established policies that allow pregnant women the option of being reassigned away from computers during their pregnancies. A few unions are already writing this option into their contracts. (This is something you and your co-workers should seriously think about if you're of childbearing age and work at computers.) In 1991, San Francisco was the first city in the country to pass an ordinance regulating VDTs in the workplace, but businesses lobbied against it so aggressively that a clause requiring computers to be spaced apart to protect workers from VDT emissions was removed. A bill has been introduced before the U. S. House of Representatives (H. R. 3528) that would require the labor department to set VDT EMF emissions standards and direct employers with more than twenty-five employees to offer pregnant women a reassignment option. The Food and Drug Administration is also testing VDT emissions.

Many people are becoming more aware of the problem. Right now, if you work at a computer, you're probably wondering whether or not you're being exposed to a dangerous level of radiation from the machine. Technologically speaking, this is quite possible. Your VDT, also called a monitor or console, is actually a little television set. It has a cathode ray tube and a tiny, powerful transformer, called a flyback transformer, that produces the image. VDTs emit a variety of ELF and VLF electromagnetic radiation: The cathode ray emits low levels of x-rays; the monitors emit pulsed EMF radiation in a variety of nonionizing frequencies, including 60 Hz; and the computer body emits various levels of electromagnetic fields. Computers emit these fields all around them, not only in the front.

We know that thousands of VDTs currently in use in the workplace emit magnetic fields higher than 3 mG. We also know that a number of studies of computer emissions have been done, and warnings about their high magnetic fields have been issued. A few malfunctioning VDTs have been investigated as a result of health complaints and were found to be emitting magnetic fields as high as 10 to 20 mG. The July, 1990, issue of Macworld (which also carried a heavy-hitting article on VDT hazards by Paul Brodeur, entitled "The Magnetic Field Menace"), measured the magnetic fields given off by a number of monitors commonly used with computers and printed the magnetic field levels. They ranged from 5 to 23 mG.

WARNING: the electricity around you may be hazardous to your health

We know that studies have found that people who are exposed to fields higher than 2.5 mG have an increased risk of cancer and other diseases. And don't forget to take into consideration what we know about the importance of the duration of exposure.

If you own a gaussmeter, bring it to work and measure around your computer. Don't forget to measure around the sides and the back, where the highest emissions are often found. You really can't predict what you'll find. Presumably, models manufactured prior to 1983 — when a few computer manufacturers like IBM and Apple began producing low-emission monitors — are likely to have high fields. But you may have an older VDT and discover that it has surprisingly low fields. If you discover high EMFs, take precautions to protect yourself and your co-workers.

There are a couple of quick, not necessarily cheap, fixes, like low-radiation monitors or the few VDT screens on the market that purport to cut down on emissions. But the effectiveness of these screens is questionable, because, as you remember, you couldn't see through anything that would actually eliminate the magnetic fields. For the most part, they probably cut down on glare and x-rays coming from the monitor. There are even a few companies that will retrofit offending VDTs to lower the magnetic fields to a safe level, for as little as $200. Safe Technologies offers such a service for a variety of models.

The best solution is simply to do the sensible thing and keep your distance — that is, three feet. Have a meeting and inform your co-workers of the problem, and they'll probably agree with the change. At that distance, no one in the room will be in danger. And remember to sit back at least two feet from your own screen. (If you have a gaussmeter, take some measurements to verify the distance at which the magnetic fields are reduced to 1 mG or less.)

The computer rooms at your children's schools should also be designed to keep the kids out of the magnetic fields. You may also try to develop the habit of turning off the computer when you're not using it. Remember, magnetic fields go away completely when a machine or appliance is turned off. Many people in offices are used to running their computers all the time, whether or not they're in use, but you could organize your work day differently so that you don't have to keep that computer running all the time.

Some large companies — among them, the New York City Board of Education and the purchasing department of the city itself — have addressed this problem by setting up purchasing standards that require them to buy low-emission VDTs. As a result of consumer concerns, computer manufacturers are already taking the magnetic field emissions/health risk problem seriously enough to spend money developing new designs to reduce fields. These same companies are also lobbying for U. S. emission standards that would duplicate the ones in Sweden.

EMFs in the Workplace

Don't forget to measure the magnetic fields around all the other business machines, too. Photocopiers are notorious for creating high fields, and many people lean on them while they're waiting for the copies to be made. It's better to press the buttons, then walk across the room to escape the field. Copying machines frequently have fields of 5 mG or higher. Then there's the fax machine to consider. And the telephone equipment, especially if your office has a switchboard. Telephone switching equipment of the sort that's used in large offices has been linked to cancer in a number of studies. Workers who use a light box to view color transparencies are being exposed to magnetic fields as high as 80 mG.

Besides office equipment, a high source of the electromagnetic fields you experience every day at work may be the actual building itself. According to Richard Tell, of Richard Tell Associates, a Nevada testing group,

> Large commercial buildings often have office areas immediately over or adjacent to electrical switch gear rooms, where the electric power comes in from the system and is switched through the building via large conductors attached to bus-bars or cables (often run through a bus duct) in the walls. If these high-voltage cables are attached to the ceiling of the basement, or make a long horizontal run beneath the floor of certain offices, as they often do, the offices directly adjacent to the switch room or these cable runs are subject to extremely high magnetic fields.

Tell has measured fields from several hundred mG to 3 G at the surface of walls, and 200 to 800 mG throughout offices that are directly over switch gear rooms. "I discovered this while working on projects that were originally driven by other concerns, say, warped VDT images or a lot of noise on the telecommunication systems," Tell explained. "If there is any interest in examining the health effects of magnetic field exposure, there's ample opportunity for epidemiology types to study the people in large office buildings. I've seen some of the EPRI studies of utility climbers (linemen), and I don't believe they were ever exposed to such high fields. As I recall, the high exposures for utility workers were somewhere in the range of 20 to 30 mG. Now here are office workers unknowingly experiencing hundreds of mG all day long." Tell believes the same situation could easily exist in large residential buildings that have their own transformers, especially if the building provides heat and cooling for its residents.

Retrofitting existing buildings would be "tough and very expensive," in Tell's opinion. If a company has enough space, reallocating the high magnetic field rooms may be a solution, but with an eye to carefully selecting what is located there. For example, moving workers out and installing a telecommunications center wouldn't be a good idea, since the fields would interfere with all the electronic equipment. Storage might be the only safe use of the room.

Shielding is also likely to be expensive and difficult. Larry Maltin, of AMUNEAL Manufacturing in Philadelphia, reports that his company did such a job for a number of businesses with similar problems, at a cost of about $70,000.

On the other hand, Tell and others think the proper place to start to resolve the problem is with new buildings — to design and build them so that the electric layout minimizes the fields:

> It's already happening in the power industry. They're designing their lines to minimize fields. It doesn't seem strange to me that builders could begin to react similarly in response to consumer demands. For example, housing design could advertise field-free homes or something like that. Make it a selling point. I think this kind of natural activity could come about as a natural response to the marketplace. Big Realtors all over the country are already becoming sensitive to the EMF issue, so far as the siting of new developments. With this, it's the same thing. And we already know how to do this. We just need to consider the practical approaches to the various methods and techniques that would lessen the fields, then build them into the designs. Right this instant, people could sit down, if they wanted to, and redesign the wiring of a building to reduce people's exposure to EMFs. And it wouldn't cost them any more. I think you could jump on the bandwagon and do it today. Right now.

"When you really come down to it," Herb Kaufman, program manager for the Empire State Electric Energy Research Corporation (ESEERCO), explained, "the problem is on the customer side of the meter. So it's the owner who has to deal with it. When you build a house or a building, you should plan the service drop to limit the magnetic fields in the building. At the point of construction, it's a no-cost item."

Writing in *Architecture* in July, 1991, Alex Wilson advises architects that, "A general goal should be to minimize magnetic fields in those areas most heavily occupied." To accomplish this, he suggests siting buildings away from electric power components and arranging interior space to minimize exposure to external sources of magnetic fields. He suggests discussing options about magnetic fields with the project electrician, electrical engineer, or utility service representative "well in advance of the design stage."

Wilson says that magnetic fields produced within a building are an even more critical concern, involving such things as: "proximity to electric wiring; structural steel or metal plumbing systems used for grounding; location of main distribution transformers and switching equipment; placement of large electric loads, such as elevator motors and HVAC (heating, ventilation, and air-conditioning) equipment; use of industrial process equipment with heavy current draw; location of office equipment, home appliances, and other electric loads."

He advises architects to "design a basement in which all incoming conductors are placed as close together as possible," to cancel fields, and to locate major electrical equipment "as far as possible from occupied spaces and on the highest floors." He also advises choosing heating systems other than electric, which are known to produce high magnetic fields.

One such sick building, as situations like this are coming to be known these days, is the building that houses the San Jose offices of Pacific Bell Telephone. In 1990, workers began to suspect the building had some sort of environmental problem, when it was found that eleven out of sixty-five people who had worked in a particular basement office developed cancer, while no one in the other floor of the building had the disease. It turned out the basement office was adjacent to the same sort of electrical switch room that Tell has described. EMFs in the basement office were found to be over 20 mG.

If you're worried about similar fields in the building where you work, first locate the switchroom — ask the manager or the building maintenance people, or look at the electrical blueprints — then measure magnetic fields in the rooms directly over the switchroom and adjacent to it. After that, you can try to trace the vertical route of the distribution cables.

ELECTRICAL WORKERS

Electrical occupations are of primary concern to EMF researchers around the world today. Study after study found that workers with high EMF exposures have a much greater risk of cancer, and new studies are being reported every month. These studies have found that electrical workers have a greater chance of developing certain kinds of leukemia and brain tumors, as well as overall cancer. (See Appendix A.)

It has also been reported that workers in certain types of electrical occupations that involve exposure to spark discharges — such as switchyard workers — may be subject to genetic damages they can pass on to their children. Children of electrical workers have more cancer and birth defects than children whose fathers don't work in electrical occupations. In addition, electrical workers' wives experience more miscarriages and increased difficulty becoming pregnant than women with husbands in other occupations.

The electrical workers we're talking about here are utility and telephone line workers, radio and telegraph operators, electrical and electronic engineers, general utility company employees, electric substation workers, switchyard workers, radio and television repairpeople, electricians, electronic equipment assemblers, ham radio operators, aluminum reduction workers, as well as certain military personnel.

A study of leukemia in navy personnel conducted at the San Diego Naval Health Research Center (Garland, *American Journal of Epidemiology*, vol. 132, no. 2, 1990) found that electricians' mates had an increased risk of leukemia and linked the finding to exposure to electromagnetic fields. The authors stated that, "This finding should be considered in the context of the literature supporting an association between exposure to electromagnetic fields and increased risk of leukemia." According to their report in the *American Journal of Epidemiology*,

> Electricians' mates operate, maintain, refit, and install ships' electrical power lines, lighting systems, and other electrical equipment. On shore, electricians' mates stand watch near generators and other major electrical equipment. Similar occupations are electricians, electrical generator operators, electric power linemen, and electric motor repairmen. Electricians' mates are frequently in contact with running equipment . . . (and) operate in an environment in which the 60-Hz electric power is generated, transmitted, and transformed.

Dr. Juan Alcedo, an environmental consultant in California who is an ex-naval officer, said he has vivid recollections of a procedure the navy carried out on its aircraft carriers, called "degaussing." "I remember when I was on active duty that, about every month or so, we'd go around and degauss the carrier. I never knew what that meant, at the time, but they did it by going around in a small boat with some sort of electronic equipment." As you might guess, degaussing was done to lower the magnetic fields that build up on such a ship, with its electric and electronic equipment, transformers, and generators. The problem with such magnetic field buildup was two-fold: Fields would interfere with the functioning of critical electronic equipment and also prevent mine-sweeping activity that was essential to the ship's security. (How the degaussing was done remains a mystery.)

Civilian utility workers perform similar tasks to electricians' mates. In particular, they're routinely exposed to running electrical equipment, and they also spend their time in an environment in which the 60-Hz electric power is generated, transmitted, and transformed. A 1988 EPRI-sponsored dosimeter study assessed utility workers' EMF exposures and found the workers were regularly exposed to 10- 30-mG fields, with peaks or spikes that went all the way up to 180 mG.

A similar study (Deadman, et al., *American Industrial Hygiene Association Journal*, vol. 49, no. 8, 1988, p. 409), carried out by McGill University's School of Occupational Health in Montreal, Canada, measured exposures of four specific types of electrical workers: transmission linemen, splicers, apparatus mechanics, and generating station (substation) operators. The study found that magnetic field exposures were seven to twenty-one times as high at work as general ambient exposure, and, as would be expected, the highest magnetic field exposures occurred while the employees were working on live lines. Apparatus electricians and splicers had the highest exposures.

The McGill study also reported that workers' exposures to high-frequency transient electric fields (HFTE) — that is, spikes — were much higher than had been expected. These spikes, or arcing fields, come about from spark discharges and switching surges or shocks from live wires. A number of studies linked these surges to possible genetic damage in electrical workers: A Swedish study (Nordstrom, *Bioelectromagnetics*, vol. 4, 1983, pp. 91-101) of workers in high-voltage substations found a decreased frequency of "normal" pregnancy outcomes — in other words, increased miscarriages and birth defects — when the father was a high-voltage switchyard worker. And a Soviet study reported chromosomal breaks in the blood of switchyard workers back in the late seventies.

Hal Nixon, a health rep with Michigan Local 223 of the Utility Workers of America, is waging a battle with Detroit Edison to protect his substation workers from dangerous on-the-job EMF exposures which he believes are the cause of the members' cancer problems. After hearing reports in the media linking EMFs and cancer, Nixon looked into the situation at Local 223 and found that one out of every 100 substation workers died of leukemia before he retired from the job. According to a union survey, sixty-five percent of the membership considered EMF exposures their greatest concern. Nixon is trying to force the company to reduce workers' exposures. During recent contract negotiations, "I asked them if they'd map the substation fields (according to the EPRI model) and confirm the physical

WARNING: the electricity around you may be hazardous to your health

barriers to keep workers away from high fields." What has been the utility response? "They turned me down flat," Nixon explains. "Said what I was asking was too expensive." The union is also considering filing some workers' compensation claims.

In 1990, the International Brotherhood of Electrical Workers (IBEW) Local 1245, near San Francisco, responded to its workers' concerns about health risks from their exposure to electromagnetic fields, by calling for a joint study of EMFs with the Pacific Gas and Electric Company. That same year, the Communication Workers of America (CWA) Local 1118, in Albany, New York, addressed a number of queries on the same subject to the New York Telephone Company and the Eastern New York Occupational Health Group. "We just came up with the idea that the electric fields that we're working in might be dangerous," explained Jack Clark, a telephone lineman who'd been with the company nearly twenty years. "We sent a letter asking about it to corporate safety. We got a response back from them that more or less told us we had nothing to worry about. But I showed it to another guy, someone who knows a little more about all this than I do, and he says the reply's a bunch of hogwash. So we sent a letter to a doctor at the Occupational Health Group and we're going to be meeting with him, to find out what he has to say."

The Communication and Electric Workers of Canada (CWC) has called for across-the-board exposure measurements for its workers. Other unions in the United States are doing the same.

Workers in the telecommunication field — especially telephone lineworkers and telephone operators, who were the subjects of the Matanowski study — are also at high risk from EMF exposure. According to Marsha Love, at the national office of the CWA, the issue of EMFs is a matter of concern right now among workers. "We have petitioned NIOSH to follow up on the Matanowski study and assess workers' exposures. We want to know the health status of our members, especially if the new technologies put our members at more risk or not."

Dave LeGrand, at the Washington headquarters of the CWA, has been involved in the EMF exposure issue for nearly twenty years. "We initiated activity about EMFs going back to the seventies, when we had NIOSH do a reproductive safety/VDT study. In the late seventies, it looked as if the issue was going to be of great importance. But then Reagan came in and cut off all the studies. No government-funded research on this has been done since then. Right now, it seems the issue is gathering steam again."

The union wants to know exactly what its workers' exposure are, and how much they are at risk. To do this, LeGrand believes the onus is on the workers themselves. "From our perspective, it's an important issue. I question the concerns of industry because of their history of doing everything they can to prevent investigation of this. They aren't interested in having public participation of anything they do. I also question the data they come up with because they don't publish the data — they just tell us what they found."

The Bell Corp. and AT&T are both hard at work measuring exposures. "As soon as we filed our request with NIOSH, their response was to go out and collect their own exposure data. Now, they have a clear philosophical belief that there's no harm from radiation from the equipment," LeGrand explains. "AT&T has a well-regarded scientist on their staff, Ron Peterson. His biases are well known; he believes there's no dangerous effects from their emissions." (Peterson has told me , "I haven't seen any convincing evidence that there's any harm from EMF exposures.")

LeGrand fears that NIOSH has been so slow in getting on this issue that they may be hard-pressed to find any of the dangerous old-fashioned electromechanical switching equipment (the newer digital equipment doesn't have the emissions that the old electric equipment did) to find out for the union what their workers' past exposures have been. It may surprise you to learn that NIOSH only has two or three people in the agency who can take EMF measurements. LeGrand attributes this problem to the fact that there's never been much governmental interest. "They don't believe there's a problem. Just recently the hand-held equipment became available. Before that, you had to get Bill Guy (a scientist from the state of Washington) in from his lab with his lab to take measurements."

For one thing, NIOSH will be measuring fields inside the company's offices for the CWA. LeGrand explains:

> Telecommunication people are in almost the same bag as electrical workers, but the data are more murky. The outside workers have a more clearly defined risk, but the office workers are of concern to us right now. We're trying to find out what that older equipment has been emitting. We're looking at smaller installations around the country that haven't gone digital. But NIOSH will have to act fast, or there won't be any old equipment left to measure. When I ask about it, the response I get from Bell Corp. is, "Go to eastern Europe or the third world countries. That's where they sent all that equipment when they went digital."

While LeGrand himself is of the opinion, "There's something going on that concerns our workers," he admits that so far as the rank and file are concerned, there's little apparent interest in the subject right now. "The majority of these people come out of a military background, and they've been brainwashed that radiation isn't harmful. They hold a clear macho belief that there's no problem. There's a lot of awareness, but they just aren't worried about it."

Early in 1991, NIOSH showed its concerns for workers' exposures to EMF by convening a workshop to review the issue, to develop a research agenda on the health effects of EMF, and to suggest methods for reducing occupational exposure. The following findings are from the published review that the workshop reported:

- Since 1982, scores of occupational epidemiological studies have reported what seemed to be a confirmation of the association between EMFs and cancer that Wertheimer had proposed, in particular, increases in leukemias, brain cancer, skin melanoma, and male breast cancer in workers with high EMF exposures. (See Appendix A for a detailed list of findings of studies of electrical workers.)
- A number of impressive, well-designed studies with reasonably large numbers of cases observed an excess of leukemia among broadly defined electrical occupations, and this excess seemed to be higher for myeloid leukemia.
- The brain is the second cancer site that has caught the attention of occupational epidemiologists. Since 1985, at least seven case control studies of brain cancer and occupational exposure to EMFs have been published. Most of these studies have shown that people in electrical-related occupations have a higher than average probability of developing the disease. At least three studies indicated the presence of a dose-response relationship between EMF exposure and brain cancer.
- Since 1990, care has been taken to study exposed groups of workers in more homogeneous occupational categories. As a consequence, more significant excesses for leukemia and brain cancer have been observed.
- The observation of male breast cancer among exposed workers may carry a particular significance in view of the proposed mechanism by which EMFs could cause cancer through interfering with the melatonin hormonal system.
- One of the most disturbing observations is (excess) brain cancer in children of exposed workers or children whose parents used electric blankets before their birth.

So far as the electrical workers themselves are concerned, there's probably a lot more overall interest in this issue than LeGrand thinks. Addressing the question of electrical workers' need-to-know, the New York City Labor Institution, funded by the New York State Department of Labor Hazard Abatement Unit, has just completed a curriculum designed to help educate groups of workers about the dangers of EMFs. This training workbook has information pertinent to workers' experiences, background data on EMFs and health, and useful examples of what other concerned workers' unions are doing.

There has recently been a flurry of union activity directed at both assessing workers' exposure and prodding the government into establishing some sort of occupational magnetic field regulations. Whereas many European countries have electric field regulations in place, only a few have attempted to regulate occupational exposures to 50- or 60-Hz magnetic fields, and the limits they've set are high enough to be useless. Germany and Poland have established 5 microtesla limits for power utility workers; and the Soviet Union limits welders to one hour a day at 7.5-microtesla fields and eight hours a day at 1.8-microtesla fields. England is proposing a 1.7-microtesla limit for power utility workers (0.1 microtesla = 1 G = 1000 mG). All of these regulations would allow workers to be exposed to fields well over 1 G, which is no protection at all, when you consider that the Matanowski study found that electrical workers at the New York Telephone Company — who had daily exposures of 4.3 mG — had seven times the risk of cancer than other workers.

Besides lobbying for EMF research and some government standards, a few unions have succeeded in addressing the EMF issue by negotiating for some forms of protection that are written directly into their contracts. As you've seen, most of these concern VDT exposures, but a few industries are starting to institute overall EMF monitoring and medical testing of exposed workers.

RF/MICROWAVE WORKERS

Perhaps the most famous case of a worker who became ill from exposure to on-the-job electromagnetic radiation was Sam Yannon, a radioman with New York Telephone. For fourteen years, until he became too disabled to work, Sam maintained a large installation of RF transmitters on the Empire State Building. He died one year before his wife, Nettie, filed her precedent-setting workers' compensation claim in 1975. (See Chapter 8.)

You may have seen TV footage of Sam before and after he was stuck down by the microwaves he spent his work life around: The radiation sickness turned a robust man in his late fifties into an emaciated skeleton in a

WARNING: the electricity around you may be hazardous to your health

matter of a few years. His doctor, Alfredo Santillo, testified that Sam had died of microwave radiation sickness. The telephone company claimed the death was not work-related. The case has had a long history of appeals and cross-appeals, but it has established some important legal precedents. In 1981, the Workers' Compensation Board ruled that Sam's death was the direct result of on-the-job microwave exposures. The next year, the New York State Supreme Court made a similar ruling. The Yannon case is currently on appeal, but it has already amply documented the fact that microwave radiation sickness can be an occupational hazard for men and women who work in an RF environment.

People who work in the RF/microwave environment include civilian and military radar technicians, civilian and military air traffic controllers, employees at radio and TV transmission facilities, and factory workers who operate RF sealers. (Microwaves are at the high end of RF radiation and the two terms are often used interchangeably.)

In the United States, RF workers have been in danger from the radiation at work for decades, because what has stood for a safety standard has been far too high to protect anyone. The RF standard in use today was established in the fifties by the military for radar workers and was subsequently adopted by the American National Standards Institute (ANSI) and then, later, by the Occupational Safety and Health Administration (OSHA). The standard is 10 mw/cm2. To give you an idea of the inadequacy of this standard, the Soviet Union keeps workers out of fields that are over 0.01 mw/cm2, which is one-thousandth of our "safety limit."

This year, a new interim RF standard has been set by the Institute of Electrical and Electronic Engineers (IEEE). Although the IEEE standard is much lower than the old ANSI standard — 1 mw/cm2 — it's still a hundred times higher than the one that's been used for years by the Russians.

The problem is that none of the American standards really mean anything in the workplace because they're all only advisory or voluntary standards that will not hold up in a court of law. As in the Yannon case, however, the legal issues can be settled by other means than measuring workers' exposures.

Much of the standard setting has been in the hands of the military, which had a vested interest in maintaining the status quo. Like industry, of course, the military had a great deal to lose if the RF standard were lowered dramatically and enforcement legislation were enacted. Besides workers' compensation claims, the Pentagon could suddenly find itself with obsolete weapons if they couldn't meet new, more stringent RF radiation standards. In general, it takes a decade or so to develop a new generation of weapons.

RF Sealers. The plight of RF sealers (the workers, not the machines) warrants immediate legislation. You may have noticed that everything you buy these days is sealed in plastic. All this plastic sealing is done by a factory worker, called an RF sealer, who spends his or her day operating an RF radiation machine that seals the plastic. These machines have routinely been measured and found to emit radiation that's even higher than the old (high) ANSI standard. Right now, thousands of RF sealer operators are currently spending about eight hours a day in fields of RF radiation higher than 10 mw/cm2.

It would cost a small amount to make these machines virtually field-free. All that is needed is a retrofit similar to the sealing of microwave oven doors to prevent them from leaking.

HAM RADIO OPERATORS

Occupational studies have linked high EMF ham exposures to higher rates of cancer. David G. Cassidy, associate editor of *73 Publications' Amateur Radio Today*, says,

> We were one of the first magazines to start publicizing the risks from electromagnetic radiation. I think the evidence is conclusive. We don't need any more studies to prove that exposing yourself to low-level radiation will kill you. Amateur radio operators have been dousing themselves in these low-level frequencies since the twenties. I think it's time they realized there's plenty of ways to protect themselves. First of all, I tell everyone to put the transmitter across the room from them. There's no reason to have it right under your nose. Once you turn it on, you have nothing else to do with it.

The magazine has carried a number of stories warning about the dangers of exposure. Cassidy and other editors routinely answer questions from readers and send them a bibliography of EMF sources. Cassidy says,

> The outdoor antenna isn't the problem. It's the high-intensity stuff coming out of the transmission equipment that's right there in the room with you. These guys are sticking their nose in this stuff and turning it on. That's what's hurting them. Some amateurs will clip a small antenna right on their shoulder when they're transmitting. Sometimes, guys in an apartment in the city will even set up the antenna inside.

Cassidy advises ham operators to get a gaussmeter and measure, then take steps to remove themselves from the high magnetic fields.

POLICE

Another special case concerns police radar guns. The strongest response to date on an occupational EMF hazard occurred this year when police departments began banning the use of hand-held radar guns. So far, departments in San Diego, California, the entire state of Connecticut, and St. Petersburg, Florida, have discontinued the devices. According to a spokesperson from Connecticut State Police, the action was taken because of concerns that troopers could develop cancer from long-term exposure to the electromagnetic radiation emitted by the devices. The move came after three troopers filed workers' compensation claim saying they had developed cancer from using the radar guns. One, an eighteen-year veteran with the Windsor Locks Police Department, said he had developed a malignant tumor on his left testicle after years of resting his radar gun between his legs. Officers in a number of other states have been filing similar claims.

Kustom Signals, Inc., of Kansas, the company that manufactures the devices, denies that the guns pose any health risk at all. A spokesperson from the company cited tests showing that emissions from the guns are well within government safety levels.

In 1990, a "California physician consulting on a thirty-two-year-old police officer with lymphoma, informed the patient that there was an increasing amount of data linking exposure to electromagnetic RF radiation to various cancers, especially leukemia, lymphoma, and cancer of the nervous system. He also pointed out that the state of California had found that police officers have a 2.68 increased risk of death from lymphatic cancer, and double the risk for all other cancers. He concluded, "When these facts are taken with clinical presentation — an enlarged lymph node in the *exact area where he rested a radar gun on his lap for hours every day*, I am inescapably drawn to the conclusion that there is clearly more than a chance relationship between (X's) disease and his chronic occupational exposure to a specific type of electromagnetic radiation that has a long history of causing biological effects."

The Federal Drug Administration recently recommended the units be kept at least six inches from any part of the body when they're in operation. Some departments are giving their officers a choice about using the guns.

Police officers are also exposed to RF radiation from their two-way radios, which are worn on their shoulders, with the antennas close to their

EMFs in the Workplace

heads. Motorcycle officers have an additional antenna right behind them on the bike, so they're sitting right in the middle of a field. Squad cars have an antenna in back, with the same result.

SAFETY FOR ALL WORKERS

With so many workers exposed to EMF health risks, what can be done? The NIOSH workshop ended with a call for industry to begin to work toward an overall reduction in workers' exposure to EMF. These suggestions are an excellent place to start. Some of the NIOSH recommendations were:

- Studies to identify and characterize EMF sources.
- Changes in the National Electric Code to reduce occupational exposures to EMF that result from dispersed neutral return currents in the ground systems of buildings.
- Designs for placement of wires in new construction that would reduce exposure.
- Development of electrical wiring installation practices that would lead to lower EMF levels in the workplace.
- Research into field cancellation techniques.
- Wiring and circuit design of industrial equipment, power tools, and office appliances to reduce EMFs, and possible retrofit of existing equipment.
- Development of more effective shielding materials.
- Administrative controls of work practices to reduce exposure, such as increasing the distance between the worker and source; minimizing exposure times; task redesign; work-station redesign; use of robotics.
- Design of lower EMF appliances.
- Modification of circuit design in equipment to reduce the generation of transient fields.
- Development of better protective equipment.
- Training and education of workers to reduce exposure.

Chapter 8

RF/MW Radiation: Towers, Radar, Cell Phones

> As technology advances and they (cell phones) get more competitive and cost less, teenagers will be using them. In their case, I'm concerned about the cumulative effect.
>
> HENRY KUES, head of the Johns
> Hopkins Applied Physics Laboratory

After fourteen years maintaining the antennas atop the Empire State Building for New York Telephone, Sam Yannon became sick with microwave poisoning (radiation sickness). Doctors agreed his illness came from exposure to the radio antennas. Yannon died horribly in 1974. By then, he was blind, deaf, mentally disabled, down from a robust 180 pounds to under 100. In TV news clips from the time, you see him in his hospital bed; at sixty, he looks like a concentration camp victim in his nineties. Sam's widow, Nettie, won a worker's comp case — the first to establish that a worker's injuries resulted from exposure to on-the-job RF radiation. "I never gave up," Nettie told an interviewer. "They wanted me to say there was no connection between Sam's illness and microwaves, but that just wasn't true."

Robert Strom has lived with his terminal chronic myelogenous leukemia for over fifteen years now, a survival rate he credits equally to medical science, his religion, and his family's love and support. In his fifties, Bob says he's especially grateful "God let me stay around to enjoy my four grandbabies." When he first learned that his work had given him cancer, he wasn't sure he'd be able to see his grandchildren growing up.

Strom was exposed to pulsed microwaves on an electromagnetic weapons project at Boeing Aircraft in Seattle. Attorney Michael Withey, of Seattle's Schroeter, Goldmark and Bender, filed a 1989 class action workers' comp suit for Strom and 700 of his fellow workers. Withey claimed the workers were "human guinea pigs. Boeing exposed them to pulse-modulated RF radiation, then monitored them to study the effects of the exposure." In 1990, Boeing settled with Strom for over $500,000 and agreed to set up an ongoing health program for the rest of the workers, who retain the right to seek compensation for work-related illnesses. (Since then, other workers have been diagnosed with cancer; some have died.) With his award, Strom established the Robert Strom EMF Foundation to educate the public about EMF hazards.

Strom and Yannon were point men in a battle between the industries that produce hazardous radiation and people who spend their workdays in these fields. Theirs were occupational exposures. But you don't have to work with microwaves to be exposed to them. In our ordinary lives, our daily comings and goings, you and I are routinely exposed to worrisome levels of the same radiation that struck these two men down.

We've had warnings about RF radiation in the environment for decades. In the early seventies, the Electromagnetic Radiation Management Advisory Council told the government we needed "research, monitoring and control (of this ambient radiation) — or in the decades ahead, man may enter an era of energy pollution of the environment comparable to the chemical pollution of today." We have already entered that era.

RF/MW RADIATION

Radiofrequency (RF) radiation is nonionizing electromagnetic radiation from 0 to 3000 gigahertz (GHz). (See the Electromagnetic Spectrum in Chapter 2.) The term "RF" is generally used to represent both radiofrequency and microwave (MW) radiation; microwaves are at the high end of the RF band.

Most people today are aware of the hazards of 60-Hz EMFs. Now we're beginning to worry about the hazards of RF/MW EMFs — thanks to the rapid growth of some of our favorite technologies, among them broadcast transmissions, the personal communications industry, the military, modern transportation, even medicine. The downside of all this technological advance is an environment awash with radiation. We're virtually surrounded by antennas, towers, microwave dishes.

WARNING: the electricity around you may be hazardous to your health

Sources of RF radiation are AM and FM radio transmissions; TV (UHF and VHF); ham radio; CB radio; cordless phones; cellular or mobile phones; microwave communications technologies; microwave ovens; police, military, air traffic control and weather radars; telephone microwave relay systems; satellite communications earth stations. Unfortunately, as with 60-Hz radiation, you don't even know it's there. By the time your body warns you, it's probably too late; we only pick up on RF effects when they've exceeded a safe level.

The closer you are to an antenna, the higher your potential exposure. Most of us are continuously bombarded by a combination of ambient RF fields from a variety of antennas. This probably has a cumulative effect. According to the EPA, which has done RF research and monitored RF levels in cities around the country, the highest public RF exposures occur near the base of broadcast towers and in high-rise residences and office buildings in line-of-sight with these powerful broadcast beams. The EPA has been raising red flags on the potential hazards of ambient RF radiation for some time. In 1986, the agency warned:

> Essentially everyone in the United States is continuously exposed to low levels of RF radiation and some people who live or work near powerful sources are exposed to higher levels. RF radiation sources have steadily increased in number and their uses have diversified so that a general increase in exposure levels in the environment has occurred. Various effects occur in experimental animals exposed to intensities found in the general environment. The increase in sources coupled with a better understanding of biological effects has heightened concerns about potential adverse effects on public health from exposure to RF radiation.

As Louis Slesin, longtime publisher of *Microwave News*, put it, "What I want to emphasize, this is not fear of the unknown, it's not somebody's fantasy. There's plenty of data out there demonstrating hazardous effects."

While the experts on both sides of this issue agree to a need for more health effects research, there's already a large body of literature reporting dangerous RF effects. Decades of studies have found chromosomal damage, eye lesions and cataracts, increased tumor growth, reduced sperm levels, stress reactions, immune system disorders, birth defects, central nervous system effects, changes in cell functions and chemical production, behavioral changes — all from exposure to RF and MW radiation. Despite the lack of laboratory studies with cancer or mortality outcomes, many well-respected epidemiological studies of workers link RF exposure to increased melanomas, brain cancer, leukemia, Hodgkin's disease, breast cancer, lymphomas, and overall cancer deaths.

Today, when RF sources and public exposures are increasing so dramatically, why don't we hear more about the potential hazards of RF EMFs? Because powerful vested interests want this decade's untrammeled technological expansion to continue in order to advance the high-ticket high-tech industries. It's all about money. With a potential for billions in profits, the industries pumping all this radiation into the environment don't want anything to get in the way of their unprecedented growth. As Nicholas Steneck pointed out in his book, *The Microwave Debate*, this stands in the way of public protection. "The main restraint against greater conservation (in RF safety standards) is a desire to maximize opportunities to expand the use of RF technology."

SAN FRANCISCO: A CASE IN POINT

"Society is enamored with new technology, but no one is talking about possible health hazards," says William Lee, ex-Chief Administrative Officer of San Francisco and one-time chief of toxic control for the City Health Department. "I think it's a major issue that needs to be looked at. The problem is, the whole telecommunications industry is being deregulated. This includes cable TV, satellite dishes, wireless, beepers, cell phones — all the new technologies. I don't think anyone knows what the effect of all this radiation is. We need research about radiation effects. How many signals are being bounced through you? TV, radio, satellites? There may be a synergetic effect. Mutagens like radiation change the DNA structure. They affect future generations. And who knows when the effects take place? What happens to children? Or teens? People worry about cancer."

When San Franciscans worry about cancer, many worry about Sutro Tower — a 977-foot broadcast tower with antennas for ten TV stations, four FM radio stations, twelve two-way paging devices, and twenty studio microwave antennas right in the heart of the city, cheek by jowl with some of its most populated neighborhoods. (Digital television antennas are soon going to be added to this array.) When the EPA advises communities to practice prudent RF radiation protection in populated areas, Sutro Tower is just the sort of facility they're warning against. Many experts told me the twenty-year-old tower would never have been built today.

In particular, Sutro worries residents of Noe/Eureka Valley, a neighborhood less than a kilometer away from the tower with a high incidence of childhood cancer. John Barry, a Realtor whose 9-year-old daughter Louise was diagnosed with acute myelogenous leukemia in 1987, persuaded the city health department to investigate the cancers after he learned that quite a few of the kids in the Noe/Eureka area had leukemia. "I knew what clusters were and I thought this was a cluster," Barry recalls.

The health department found a cancer cluster — nearly twice the expected rate of childhood cancers in Noe/Eureka Valley children — but failed to link the disease to EMFs from Sutro or to any other environmental hazard. Faced with public opposition and the damning health department study, Sutro Tower planners withdrew a pending application for expansion.

Then, in 1990, Steve Selvin, a researcher at the University of California at Berkeley School of Public Health, conducted his own study and found that "pediatric cancers occurred closer to one another (clustered) and to Sutro Tower than would be expected by chance."

While some city officials and the people at Sutro claim Selvin merely used Sutro Tower as a "geographic reference," Selvin's report clearly states: "An environmental exposure potentially associated with occurrences of childhood cancer is electromagnetic radiation emitted from a large microwave tower located a few kilometers SW of the center of the city." Selvin defined "exposed" individuals as "all those under 20 years of age residing within 3.5 kilometers of the microwave tower (Sutro)."

Selvin's work caught the attention of Dr. Raymond R. Neutra, chief of Special Epidemiological Studies Programs for the State Department of Health Services. Dr. Neutra, a major player in California's EMF research, wanted to do a follow-up on Selvin's work — adding magnetic field measurements at case and control homes. "We got together with the EPA and wrote a proposal based on Selvin's work, combining topography with frequencies," says Neutra. "But we couldn't get funding. I'd still like to do it. And I think this should be looked at in more than one city."

Neutra recalls that, when the EPA measured RF fields in San Francisco in the seventies, "When you couldn't see Sutro Tower, the fields really went down." If you could see Sutro Tower from your house, you were probably getting zapped. Neutra cites studies in Honolulu, Hawaii; Sidney, Australia; and Portland, Oregon, that found more cancer cases in census tracts with transmission sources than in tracts without them. (The fact that a Honolulu census tract is 1.5 miles in area does not bode well for the ten schools less than a mile away from Sutro.)

In 1995, when the San Francisco Unified School District wanted to expand Rooftop School, a school directly in line with Sutro, parents worried about radiation from Sutro and opposed the plan. Richard Lee, of the health department, told them RF levels there would be higher than ambient levels elsewhere in the city — due to Sutro Tower. Lee says, "Based on what I know, they'd be two or three times higher. The EMFs are higher the closer you get to the tower. We took measurements in '73 and '74. The highest EMFs we found were at Rooftop, Noe Valley, and U.C. Medical Center."

RF/MW SAFETY REGULATIONS

In the eighties, the City of San Francisco hired the firm of Hammett and Edison to measure Sutro's emissions; H&E reported that levels fell safely below the RF ANSI standard. In 1993, H&E remeasured the fields. Eugene Zastrow, present manager of Sutro Tower — who agrees that "in the best of all possible worlds, all the antennas would be far away from people" — told me the fields from Sutro still "fell below the ANSI standard."

Whether that meant they were safe is another question.

W. Ross Adey, a prominent physician at the Jerry L. Pettis Memorial Veterans Hospital in Loma Linda, California, who has spent nearly three decades on EMF research, didn't think so. In 1990, at Dr. Neutra's request, he advised San Francisco that:

> " . . . with respect to the unfortunate use of ANSI standard C95.1-1982 as a benchmark criterion, even though this has been accepted by the FCC as its guideline to which adherence is required by all broadcast stations at this time ... the standard was prepared primarily by engineers who did not recognize even the existence of athermal field effects. . . . The ANSI standard has become an archaism in virtually all the rest of the world. . . . "

Many countries, including Great Britain, Japan, Germany, and Russia, have set up much more stringent RF public exposure standards. And the World Health Organization (WHO) recommends an exposure limit five times as strict as ANSI and advised that "exposure of the general population should be kept as low as readily achievable."

But, for decades, what passed for protection in this country was the ANSI/IEEE standard C95.1-1982. (American National Standards Institute/Institute of Electrical and Electronic Engineers.) Revised in 1992, there were still many problems with ANSI — enough for the EPA to come out against FCC adoption of the standard.

In a 1993 letter to the FCC, Margo T. Oge, director of the EPA Office of Radiation and Indoor Air, wrote: "EPA recommends against adopting the 1992 ANSI/IEEE standard because it has serious flaws that call into question whether its proposed use is sufficiently protective of public health and safety. One of the problems Dr. Oge had with ANSI was that it's based entirely on thermal effects (the heating of tissues) and ignores a growing number of studies reporting harmful effects at nonthermal levels.

ANSI is basically an industry standard, not a public health protection standard at all. To make matters worse, it's a voluntary limit — not admissible in court. The ANSI standard is a dinosaur, a palliative for industry. A company can address its workers' concerns by assuring them that radiation levels are in line with the ANSI limits. And many a community has been lulled into a false sense of security by experts who tell them the RF field they're worried about "falls within the ANSI limits."

The RF safety limits recommended by the National Council on Radiation Protection and Measurements (NCRP) — charged in the eighties with developing national 60-Hz and RF radiation safety standards — are more protective. In 1996, NCRP recommended RF exposure limits five times more stringent than ANSI. NCRP also set separate limits for people working around RF and for the public. (Safety standards typically allow higher exposures for workers than for the general public.)

NCRP's limits began at nonthermal levels and took into consideration differences in effects from continuous wave and modulated radiation — like the fields from today's cellular and wireless technologies. Tom Tenforte, director of the NCRP radiation project, said that with the "tremendous growth of the new wireless technologies" it was important to take a look at the possible health effects of modulated MW radiation.

In 1996, the Federal Communications Commission (FCC) bowed to the EPA, over heavy industry lobbying, and adopted RF health and safety regulations based mainly on the NCRP recommendations (and in part on ANSI) "to protect workers and the general public from potentially harmful RF emissions due to FCC-regulated transmitters." According to Dr. Robert Cleveland, director of the FCC Office of Engineering and Technology, "We had 2,800 pages of comments (on ANSI) before making our decision. The way it broke down was, most industry supported us using ANSI, while the federal agencies had some reservations."

The 1996 FCC RF exposure limits are much more stringent than ANSI: 1 mw/cm2 for public exposure and 5 mw/cm2 for workers. Milliwatts per centimeter squared (mw/cm2) is the unit used to measure the intensity or power of energy. The Specific Absorption Rate (SAR), also expressed in mw/cm2, is a measure of the amount of energy absorbed by a body in an RF field.

The FCC regs now require cell phones to be tested before they are marketed — under ANSI, they were exempt — and user SARs will have to be kept below 1.6 w/Kg. Today's FCC limits are more stringent than ANSI, but whether they're protective enough is still under debate; for one thing, the

FCC still bases its RF standards on thermal effects.

The FCC refused to mandate federal preemption of state and local zoning decisions regarding cellular towers — something the industry had lobbied long and hard for. The same was true for the Telecommunications Act of 1996; the act expressly preserves city and state zoning authority. (Interestingly, cellular companies faced with resistance from local communities routinely inform them that they're required by the act to establish wireless capabilities within the year — this is not true.) Which brings us to the issue of the safety of cell phones.

THE CELL PHONE DEBATE

When they hear the word "microwaves," people generally think: microwave ovens. Microwave ovens have been regulated since the fifties, when *Consumer Reports* found they were leaking hazardous levels of microwaves, and the government required manufacturers to improve door seals to eliminate leakage. (If you have a microwave oven you should still be careful. Stay a safe distance — at least five feet away while it's in operation — and have it inspected once a year for leakage. Especially, keep children away from the microwave. They love to put their noses up to the glass and watch things cook so magically — in the process, they may be cooking themselves.)

Today's microwave oven is the cellular (or wireless) phone, and the debate over cell phones — whether they cause brain tumors? if siting the antennas (base stations) in populated areas is dangerous? — has really heated up in the past few years.

According to Mike Houton of the Cellular Technology Industry Association (CTIA), all the furor is "just because some guy in Florida sued Motorola for his wife's brain tumor. Then he went on the 'Larry King Show' and suddenly everyone was in an uproar. A judge threw Reynard out, said it was based on junk science and talk shows."

Reynard v. NEC, which was filed in Florida in 1993 and charged that David Reynard's wife died of a brain tumor caused by her cell phone, was dismissed in 1995. David Reynard decided he couldn't afford to fight the industry and dropped the appeal, although his attorney, John Lloyd, says he feels they could have prevailed. Since then, a score of similar lawsuits have been filed, each claiming that cell phones were the cause of the plaintiffs' brain tumors. Particularly worrisome are a number of "insider suits" whose plaintiffs are former industry employees.

Robert Kane v. Motorola, filed in 1995, was brought by a twenty-year Motorola research engineer who claims his brain tumor was caused by RF radiation from the antenna on a prototype phone he field tested in the eighties. The tumor location corresponds exactly to the antenna on the phone. In Wright v. Motorola, industry employee Debra K. Wright is suing Motorola and others for brain tumors she alleges were caused by cell phones she used during her years of employment with US West Cellular and Bell Atlantic Mobile in Arizona.

These suits further allege that, although Motorola has repeatedly gone on record defending the safety of cell phones — in 1993, company spokesman Ed Staiano told the media that thousands of studies had found cell phones safe — the company failed to warn users that it hadn't conducted adequate tests establishing whether the use of cellular phones posed any risks.

How safe are the phones? That's a critical question, when you consider that there are currently forty-five million users in this country alone. In fact, to date there is no research, none, that has proven the safety of cell phones.

Early on, the industry turned to Dr. Om Ghandi, a University of Utah researcher, to give its instruments his seal of approval. Dr. Ghandi tested various models and reported, late in 1994, that levels of energy absorption in ears, heads, and brains of cell phone users were safe. His press release got enormous media coverage; it stated, in part, that "exposures from handheld cellular telephones are well within national safety standards." (What standards that would be is not clear, since ANSI had no safety limit for the instruments.)

Then, in 1996, Dr. Ghandi suddenly retracted his earlier findings, stating in another press release that SARs he found were too low and he had "revised his figures on the amount of radiation absorbed by the human head upward," owing to errors in his original calculations. Dr. Ghandi got higher SARs because he redesigned the models and placed the phones against the ear, with antennas closer to the brain — as they would be in actual use. (Also interesting is the fact that Dr. Ghandi's retraction hardly received any media attention at all.)

For years, respected scientists the world over have been quite outspoken about the possible dangers of RF/MW radiation.

A classic fifteen-year study by Dr. Stanislaw Szmigielski, a professor of pathology at Warsaw's Military Institute of Hygiene and Epidemiology, found that Polish military officers exposed to RF/MW radiation — within the accepted safety limits — had double the overall risk for cancer, six to eight times the incidence of leukemia and lymphomas, as well as chromosomal breaks. Szmigielski's subjects also had headaches, high blood pressure, memory loss, and brain damage.

In 1987, Dr. Terry L. Thomas conducted an extensive study of American workers exposed to RF/MW radiation and reported a significant increase in brain tumors. The Thomas study found a dose/response rate: the longer the exposure, the higher the risk.

Soma-Sarkar, a scientist at the Delhi Institute of Nuclear Medicine and Allied Sciences with an international reputation for his radiation work, exposed mice to microwaves — at a power density that has repeatedly been set as a safe limit for general public exposure — and reported mutagenic, canceragenic effects.

In 1990, Dr. Stephen Cleary at Virginia Commonwealth University found that human brain cancer cells proliferated rapidly when exposed to MWs — and even when they were removed from the RF fields, the tumor cells kept growing.

Henry Kues, at Johns Hopkins' Applied Physics Laboratory, exposed primates' eyes to nonthermal levels of pulsed microwaves similar to those from cell phone technology and reported "irreparable damage." Kues found that pulsed MWs caused more damage, and the application of a common glaucoma drug caused damage at a lower frequency. "When we applied a common glaucoma drug, it brought the threshold for the effect way down, below both ANSI and NCRP standards," Kues explained. (As a result of Kues' work, the Applied Science Lab set exposure standards that are 100 times more stringent than ANSI.)

Kues said he worries about young people using cell phones. "As technology advances and they get more competitive and cost less, teenagers will be using them. In their case, I'm concerned about the cumulative effect." (A later study Dr. Kues completed for Hewlett-Packard Co. that exposed rabbits' eyes to higher-frequency MWs, which H-P was using in one of their wireless projects, reported no damage. Kues plans to repeat the study using primates.)

A colleague of Dr. Kues, Dr. Henry Lai, from the University of Washington at Seattle, says the industry "has really taken pot shots at Kues" — but in 1994, work by Dr. Lai and Narenda Singh caused enough concern for the industry to initiate a large effort in damage control. The Lai/Singh study found DNA breaks in the brains of rats that had been exposed to pulsed microwaves for two hours. DNA breaks are considered a precursor to cancer.

"People can be exposed to this kind of radiation and higher from cell phones," explained Dr. Lai. "Animals were exposed for two hours, but people use their phones intermittently. This has certainly raised some red flags on cell phone use."

WARNING: the electricity around you may be hazardous to your health

In a series of December 1994 memos leaked to *Microwave News*, Norm Sandler of Motorola set out to "put a damper" on the Lai/Singh work. One memo states: "The Lai/Singh research . . . could encourage scientific speculation about a link between RF exposure and carcinogenesis . . . at a minimum . . . be interpreted as further evidence that RF exposure, even at low levels, poses at least a suspected health risk." (See Chapter 3: Controversy and Cover-up, for more.) In the end, the flack catchers at Motorola thought they had "sufficiently war-gamed the Lai/Singh issue" — but they had no idea of the studies still to come.

In one of the most compelling, Australian Dr. Michael Repacholi — head of the EMF research program at the World Health Organization (WHO) — caused a furor when he reported thermal effects and a clear cancer risk from microwave radiation of the type emitted by mobile phones. (Dr. Repacholi also raised eyebrows when evidence surfaced that his findings had been suppressed for nearly two years.) Dr. Repacholi found that long-term exposure to pulse-modulated RF radiation caused double the incidence of lymphomas in mice — and sped up development of the disease.

Another Australian, Dr. Bruce Hocking, once medical director of the Australian state telephone system, Telstra, studied a group of mobile phone users and reported that they experienced headaches, dizziness, nausea, and blurred vision. Dr. Hocking called for restrictions on mobile phone ads and recommended keeping children from using the devices.

A Belgian study (Maes et al.) reported a mutagenic effect when human white blood cells were exposed to signals emitted by a Global System for Mobile Communications (GSM) base station antenna. The general public wouldn't be likely to experience this exposure — but workers could. GSM digital systems are widely used in Europe, and a GSM mobile network has been recently introduced in the U.S. by a consortium of wireless companies — among them U.S. West, AirTouch, American Telephone and Telegraph, and Bell South — who plan to operate it nationwide. Many cellular systems operating in the U.S. are analog, not digital; a 1997 study by Dr. Ross Adey, reported no effects in rats exposed to analog RF radiation.

Other studies around the world have reported increases in lymphomas, skin cancer, breast cancer, and brain cancers from exposure to modulated RF/MW radiation.

In response to the uproar from the lawsuits and then the studies, the cellular industry — "I think our industry is acting responsibly," says Houton. "If not from Day One, then from Day Two or Three." — set up a five-year, $25 million research project, the Wireless Technology Research Group (WTRG) under its Scientific Advisory Group (SAG). SAG's mission:

to produce research about whether cell phones are a health risk. So far, four years and $17 million later, the research group hasn't done any studies except one small study that found cell phones can interfere with pacemakers and warned pacemaker users to keep the phones away from their chests. Studies have also found that they interfere with hearing aids and medical devices. In '94, the Dutch National Health Inspection prohibited the use of GSM phones on hospital premises.

Dr. George Carlo, head of SAG, explains that the group is busily preparing to do the critical studies. Meanwhile, WTRG/SAG is reportedly experiencing both financial and political troubles.

Motorola is sponsoring the largest cell phone research program in the world, with an agenda that includes in vivo, in vitro, and epidemiological studies on cancer, brain tumors, and hormones, as well as dosimetry studies. Among these are replications of some of the controversial — from the point of view of the industry — research, like the Lai and Repacholi work. Meanwhile, Dr. Robert Morgan studied 60,000 Motorola employees and reported an unexpectedly low rate of cancer mortality.

Critics of both research efforts point to problems inherent in industry-funded research and call for a committed RF/MW research agenda on the part of the government. At present, that isn't likely, as funding for federal EMF research on the whole is rapidly disappearing.

So, thus far, there is no research the industry can point to to convince us their phones are safe. When I asked Mike Houton about this back in 1995, he told me, "We have a certification process where manufacturers can voluntarily submit their phones for testing to make sure they meet the standards set by the FCC. They get a certificate. You'll see it when you go out shopping, it means they've met the FCC mandatory standards." At the time, there were no FCC cell phone standards at all.

Should you use a cell phone? The ultimate decision rests with you. With so little evidence available, do you want to be a guinea pig for an industry that has all but ignored its mandates where users' health is concerned? Many of the experts I interviewed — among them Tom Tenforte, Norm Hankin, and Raymond Neutra — are taking a conservative approach, waiting until the phones are proven safe before they get in the habit of using them.

As Hankin explained, "On a personal basis, I don't feel like exposing myself and my wife to what is the greatest source of RF exposure in the environment today. We have a car phone with a pack (a design where the antenna is away from the user). We make a couple of calls a month on it — it's in the car for emergencies. So far as base stations, the industry is talking about 100-200-foot towers, the exposure from them isn't much. I'm more concerned about the hand-held units. People who're holding these antennas right up to their heads are being exposed to more radiation. It's an area with scientific uncertainty. Why should we be exposed? The old dosimetry model is inadequate, it doesn't really tell you how modulation can be factored into dose."

As to the base stations, communities around the country are digging in their heels, determined to stop the towers' relentless march into everyone's backyard. In a way, the cell phone industry is its own worst enemy. In its enormous push to provide customers with seamless service — the industry projects some 150,000 new base stations across the country by 2000 — it has generated a wall of resistance from people worried about possible health hazards from all this new radiation. Having had to ante up billions for construction, start-up costs, and government licenses, the giant communication companies are determined to have their way — despite many communities' unwillingness to follow them where no man has ever gone before.

CTIA (Cellular Technology Industries Association) tried — and failed — to get language into the 1996 Telecommunications Act that would allow them to automatically override local opposition and zoning. What the act does say is that states or local government shall not prohibit the provision of personal wireless services and that they shall act on any request to site personal wireless service facilities within a reasonable period of time. This has been construed by the companies to mean that local governments must let them set up antennas wherever and whenever they want to. Meanwhile, these issues are being decided before municipal boards and in the courts. (See Chapter 1 for more on this.)

In the courts, according to attorney Bruce Greenberg, writing in the New Jersey Law Journal, the law has "consistently sided with the companies," finding that the towers are beneficial and may not be denied because of residents' health or aesthetic concerns.

Moreover, it seems the industry is content to do whatever is necessary to get its towers in place, even acting at times in flagrant violation of the law. When the California Public Utilities Commission (PUC) decided to find out whether cellular companies were following the rules, it discovered that many companies were routinely building towers without the necessary permits — and lying about it to the PUC. The PUC suspected that the problem was widespread.

The fact that people don't want cell phone towers on school properties or in their residential neighborhoods presents a major obstacle for the companies. Many people still believe the towers are an eyesore and are afraid they might be dangerous. When San Francisco confronted this issue in '93, over an antenna on a school building, Dr. Neutra advised: "Why would a responsible public agency go out of its way to expose children to an agent that wasn't necessary for their education and hadn't been thoroughly studied?" Leland Yee, building and grounds director for the school district, told the *San Francisco Chronicle*, "We came to the conclusion that microwave towers aren't good for the kids." The city passed an ordinance banning mobile communication towers from school property.

Asked about this, Dr. Lai said, "I get calls about this from all over the world. The base station radiation is very small, I don't think there's much danger. But I think they should not be in an area where children are. Children are more susceptible to radiation."

Despite strong opposition from the community, San Francisco hasn't been able to stem the tide of antennas throughout the city. After a years' long struggle with Pac Bell — the company wanted to place 1,500 base stations in the San Francisco Bay area — all the city could do was slow the process down by requiring separate conditional use permits for each antenna. "But the planning commission seems to always allow the permit," explained Supervisor Sue Bierman's aide June Guttfliesh, who led the charge against the towers. "People appeal the decision, then it comes before the Board of Supervisors. But we haven't stopped them."

And so "Stealth Antennas," as *The New York Times* dubbed them in a March 1997 article, are going up everywhere; companies pay property owners thousands of dollars a year to place them on their buildings in order to meet the exploding demand for wireless services. The article goes on to describe all the ingenious ways these antennas are being camouflaged to blend into the architecture or to appear to be what they are not — church steeples and even trees — in response to growing health and aesthetic concerns of the people who have to live around them.

RADAR POLLUTION

If you live near an airport or a military base, you may have more to worry about than the nuisance of noise from the low-flying planes — radar pollution. Radar is regulated by ANSI (which relies on a thermal standard developed about thirty years ago to protect workers and uses average radiated power as a measurement, rather than high peak power — which can be 500 times the average power and happens to be the point at which bioeffects occur.) Hardly anything has been done officially to evaluate the effects of radar exposure on the general public, although scores of studies of people who work with radar have shown dangerous effects. (The Szmigielski study found RF/MW exposed soldiers had much greater cancer risks, and a large study by the MIT radiation lab reported increases in Hodgkin's disease for men who were occupationally exposed to radar.)

WARNING: the electricity around you may be hazardous to your health

Unfortunately, an informal epi-study with human subjects has been going on for nearly three decades in Brevard County, Florida, in a community called South Patrick Shores. South Patrick Shores is right next to Patrick Air Force Base; some homes are less than 1,000 yards from a complex of radar domes at one corner of the base. The radar is operated by the FAA and has a 250-mile range. Residents directly in line with the powerful beam have been bombarded by a continuous level of radar since the sixties.

In 1993, the state Health and Rehabilitative Services (HRS) reported a "striking cluster of lymphatic cancer (Hodgkin's disease) in the corner of the subdivision closest to the radar." There were also elevated levels of breast and cervical cancers. Then, two years later, a cluster of amyotrophic lateral sclerosis (ALS or Lou Gehrig's disease) a deadly immune system disease, was discovered.

According to *Florida Today*, people were originally alerted to the radar beam by their appliances: TVs with interference, VCRs that couldn't be programmed, electric garage door openers with minds of their own. Even the computer system at a nearby hospital malfunctioned. Then came the news about the clusters. Residents, including Dr. Horst A. Poehler, a scientist who worked on radar at Kennedy Space Center and resides in South Patrick Shores, thought the radars had caused the diseases. "They're too powerful and too close to these residences," Dr. Poehler said.

Dr. Milton Zaret, a well-known microwave researcher, agreed. "The most obvious and immediate source (of the cancers) has to be the radar." Sam Milham, Jr., a retired epidemiologist from the Washington State Department of Public Health who has done many studies of RF radiation, said the incidences of disease would never have fallen so close to the radars through chance.

Ed Mantiply, an environmental scientist at the EPA's National Air and Radiation Lab, thinks the cause was "probably the radar." Mantiply was sent to the Shores to take measurements of the fields. He recalls, "It was really sad to go to the meetings and see all these people in wheel chairs." Mantiply, who has measured radiation in many cities in the U.S., told me, "I'd like to see epi-studies around high power AM broadcast antennas and high peak power radars. What you have to do in a case like this is look at the other radars in the country to see if there are higher disease rates around them, also."

In 1992, the FAA announced plans to close the radar facility at Patrick Air Force Base — although it was careful to explain this wasn't being done in response to the cancer scare. Both the Air Force and the FAA have gone on record stating that the radar is safe. It took five years to actually dismantle the radars and move them "out to the boonies," as a newspaper reporter put it, where no one would be exposed.

Police radar guns made the news when officers around the country started developing cancers — in the exact spot where they routinely rested the devices. If they held them in their laps, they got testicular cancer. If they leaned them against a partially opened window next to their heads, brain tumors. Under their arms, lymphomas. While the manufacturers continue to deny the devices pose any health hazard, departments around the country have banned hand-held radar guns, and a number of lawsuits have been filed by officers with cancer.

Gary Poynter, an ex-Ohio state trooper, was the first to realize "something was going on with the radar guns." He began compiling a list of cops with cancer and going around the country, speaking at meetings of lawmen. Poynter told me recently, "Right now, there's 268 names on the list."

Despite the anecdotal evidence, no studies have linked radar guns to cancer; the state of California did report that police officers are 2.68 times ore likely to die from lymphoma and have double the risk for all other cancers. By 1997, with lawsuit after lawsuit being dismissed by judges who cite the dearth of health effects science, most of the remaining suits have been dropped.

HAM RADIOS

In the early eighties, Dr. Sam Milham did two studies of amateur radio operators and reported that he found double the incidence of leukemia in the "hams." David G. Cassidy, the editor of *Amateur Radio Today*, says many occupational studies have linked high EMFs from their radios to increased cancer rates among ham radio operators. "I think the evidence is conclusive," says Cassidy. "Amateur radio operators have been dousing themselves in these low-level frequencies since the twenties. It's time they realized there's plenty of ways to protect themselves. Like — put the transmitter across the room from you. There's no reason to have it right under your nose. The outdoor antenna isn't the problem. It's the high-intensity stuff coming out of the transmission equipment right in the room with you. These guys are sticking their nose in this stuff and turning it on. That's what's hurting them."

WHAT CAN YOU DO?

What can you do about RF hazards in your community?

One thing would be to borrow the strategies suggested for grass roots 60-Hz organizing. If your concerns focus around a cell phone tower, the example of Medina (in Chapter 1) should be useful. To bolster your case with your neighbors and the city fathers, and to work out a doable mitiga-

WARNING: the electricity around you may be hazardous to your health

tion plan, it can be useful to hire a consultant firm like Sage Associates (Montecito, California) to evaluate the situation while it's still in the planning stages. Sage Associates specializes in consulting between companies who wish to install RF or 60-Hz facilities and the decision makers.

Cindy Sage explained, "One problem is, there are very few RF meters around and they're all in the hands of the industry. What you have to do if, for instance, a cell tower is proposed, is ask your local jurisdiction to require that the applicant tell you what levels of RF can be expected in your neighborhood — at the nearest home, the nearest school, and so on. They can do modeling on the computer, the data is available. Then you will need an objective and independent view of these field levels, so the decision makers can see where the levels fall and what they mean, not only in terms of existing standards, but with respect to the evolving science of health effects. As with any environmental constraint, it's advisable that you find a qualified third party to evaluate that information. Otherwise, you'll have AirTouch telling you what it means."

And then, follow the lead of people around the country who have successfully influenced decisions about 60-Hz or RF sources. Persistence, patience — especially with your neighbors who may have to be brought up to speed on the hazards — and a dossier of clear, compelling information are the tools that will help you and your community in the fight.

On the other hand, if your community discovers an antenna that's exposing people to harmful RF radiation, the 'fix' does not have to be extreme. Marin County, California, provides a good example of this. In 1995, it was brought to the attention of the County Commission that a cluster of county emergency antennas on the roof of a fire lookout in a remote area of the county was exposing workers in the lookout to hazardously high RF radiation — even in terms of ANSI. All that was required to remedy the situation and get the people out of the fields was to raise the tower — and the antennas — 35 feet. Cost to the county was about $25,000. (It is to their credit that the Marin County Commissioners reacted so promptly once the hazard was discovered.)

Epilogue

> *From my point of view, looking at it from what's happening in occupational health, there've been traditional ways of looking at health risks from nonionizing radiation. They said there shouldn't be any health risks. We're in the midst of a paradigm shift right now. Exactly what it'll be is a major research puzzle.*
>
> JOSEPH BOWMAN, NIOSH

> *Today, with the cutback in federal funding, researchers are being funded by vested interest groups like EPRI. These researchers are more inclined because of subtle pressures to only represent negative results when it comes to EMFs, or not to report positive results. They're afraid to say there's a problem and the EPA ought to regulate it. They'd be painted into a corner and have difficulty getting funded. That's why discussion on EMFs in scientific circles is so muted today.*
>
> David Bayliss, HQ, EPA EMF Group

It appears that the solution to the EMF problem isn't going to come from the government or from private industry — it's going to come from the people themselves, acting in concert to demand public health regulation and changes in energy policy. Besides making prudent changes in the way you and your family use electricity, you must get together with other like-minded people and make sure your voices are heard.

WARNING: the electricity around you may be hazardous to your health

Today, there's no way for the power companies to continue to sidestep the EMF issue. It's on too many people's minds. You've already seen how citizens concerned about EMFs have organized to attract the attention of the media and their public officials. And it's not just a question of numbers: Your success will mainly depend on smart organizing and, above all, persistence. If you identify the EMF problem clearly, target your objectives carefully, then make your demands known to the people who can respond to them — your Public Utilities Commissions (also called Public Service Commissions), your town boards, and the city council — then you're very likely to win on an EMF issue in today's arena. And bear in mind that the boards that regulate utilities are designed to hear citizens' complaints: They will respond to your EMF concerns because that is their mission.

Another avenue to take is to address your complaints directly to the utility, whether it's privately owned or public. In the case of a public utility, organize a write-in campaign where customers will include a statement of concern about EMFs with their electric bill payments. And don't hesitate to present your case at utility meetings — they're open to the public. Private utilities, which are also regulated by the PUC, are directly responsible to their shareholders, via their board of directors, to provide a safe product. You can become a shareholder by purchasing a single share in the company, which then opens a number of strategies: (1) You can speak at board meetings; (2) you can request a list of stockholders and petition them to join you in your campaign; (3) if you're able to raise enough support, your concerns can become a *proxy item*, which means that all the stockholders will be polled through the mail about it, and it will become an agenda item for discussion by the board of directors.

In the case of existing transmission lines or substations, lobby for reduced public exposure to their fields, with children's facilities given priority. According to M. Granger Morgan and Indira Nair, the following are a few possibilities to keep people out of power-line fields:

- Attempt to route new transmission lines so they avoid people.
- Widen transmission line right-of-ways.
- Develop designs for distribution systems, including new grounding procedures, which minimize the associated fields.

In situations where the power company wants to erect new power lines, focus your opposition on asking the company to prove the need for the lines. The days are over when electric companies are just given carte blanche to build a new line or increase the capacity of existing lines to meet increased customer needs for power.

Epilogue

Today, the question of need must be addressed within the context of energy conservation. Energy conservation policy holds the key to an immediate, sensible, inexpensive, and safe resolution of the EMF public health question. Energy conservation is also the answer to everyone's concern about our continued ability, as a society, to produce all the energy we need.

The time is ripe for sweeping changes in energy policy. These changes have mainly to do with something the industry calls *demand-side management* (DSM) — that is, policies at the user end of the electrical delivery system that encourage customers to conserve electricity in order to limit the need for new energy sources. By encouraging energy conservation, or energy efficiency, utilities can provide more power to their customers without having to build new generating plants or new transmission lines.

The concept of better DSM has already begun to take hold in communities around the country. In many states, the PUCs are requiring utilities to submit DSM plans that show they're encouraging energy conservation by providing financial incentives, in the form of rebates on customers' electric bills, so people will invest in more energy-efficient equipment.

The Sanford Estes home in Charleston, South Carolina is an example of how this works. Following the guidelines of South Carolina Electric and Gas' enlightened DSM policy, which offered generous rebates and the promise of lower electric bills, the Estes built an energy-efficient house and received a "good sense" rating from the utility. Last December, the electric bill for their 5,800-square-foot home was only $180. The prior year, the electric bills for their old 2,000-square-foot house were as high as $350 a month in the winter. Besides the large individual savings, you can easily see the overall benefit to the system from increased energy efficiency. Energy-efficient improvements that are underwritten by the electric company, in the form of cash rebates, can also be applied to existing homes.

Geoffrey Crandall, of MSB Energy Associates, in Middletown, Wisconsin, is an energy efficiency expert who's in the business of advising utilities and their customers on how to use energy "more intelligently at the point of end use." Crandall says, "We ask the question, 'Is this new line really necessary?' We've found that if the regulators are interested in this, the utilities are interested. In areas like California, where there are strong environmental groups lobbying for energy conservation, the utilities are practicing better DSM."

Crandall provides an example:

> We help the customer examine how energy is being used at their building, so we can reduce their need to have the power company deliver more electricity. For instance, we had a large

WARNING: the electricity around you may be hazardous to your health

industrial customer in the Midwest who was considering putting in a big energy-using metal melt furnace. The utility would have had to update their transmission line to provide electricity for the equipment. Instead, the electric company gave them a rebate to use a different kind of furnace, with a new design so they didn't have to upgrade the line with the accompanying increased fields, harmonics, and transients that people are worried about.

This can work in a home, too. Crandall explains,

> There are at least 300 or 400 different ways to conserve energy in a domicile. We ask clients, "Aren't there other options you could pursue to reduce the need for new transmission lines?" For instance, a customer can put in new high-efficiency lighting systems to displace the old-fashioned, high-energy kind with 100-watt bulbs. There are high-efficiency furnaces. Or, instead of whole-house air conditioning, they can have trees, white roofs, attic exhaust fans to draw out the heat from the house. There are all kinds of variable high-efficiency motors on the market these days.

The more consumers demand these energy-efficient products, the more manufacturers are gong to make them.

Improved DSM, in the residential sector alone, can result in huge savings in energy costs, as well as reduced exposure to power-line EMFs, since it has the effect of reducing current on the lines. Keith Laughlin, staff director of the Subcommittee on Investigation and Oversight for the Committee on Science, Space, and Technology in the U.S. House of Representatives, says your lifestyle doesn't have to change drastically, either. "You still go about your business in the same way," Laughlin explains. "There's no sacrifice involved; you don't give up anything. You just use the energy more efficiently. There are many new technologies that are more energy-efficient and the utilities have to make it attractive to customers to invest in these products."

A utility that's on the cutting edge of this trend toward energy efficiency is Pacific Gas and Electric in California. In a recent report to the PUC, PG&E said they needed 3,300 megawatts of additional electricity to meet customers' increased needs over the coming decade, and that 2,500 megawatts of this — a full three-quarters of the demand — would be coming from increased demand-side efficiency! This would preclude the need to

Epilogue

build new or larger or additional generating plants and transmission lines and save PG&E's customers a lot of money, because the cost of any upgrading done by a utility is automatically passed on to its rate payers.

There's really no excuse for an electric utility in the nineties not to be aggressively engaged in pursuing the energy conservation option instead of building more power lines. However, an electric company's motivation for building new lines may have very little to do with meeting its customers' energy needs — and a lot to do with making money.

In a 1991 hearing on EMFs and high-voltage power lines that focused on a proposed transmission line in Michigan, Representative Howard Wolpe (D/Mich.), chairman of the Subcommittee on Investigations and Oversight of the Science, Space and Technology Committee, testified:

> I began this inquiry with two admitted biases. First, I believed that even if the potential for risk to public health exists, reasonable efforts must be taken to avoid that risk. And second, I began this inquiry with a measure of sympathy with the position, as I understood it at that point, of Consumers Power Company. The generation and delivery of electricity is fraught with potential health and environmental risks. These risks must be balanced with the utility's responsibility to provide service to its rate payers at lowest possible cost. . . . I was prepared to support without much question the construction of this power line with one crucial point to be demonstrated. I wanted the proof that the building of the power line and any potential associated risks could not be avoided. I wanted to be sure that all other less risky but cost-effective means of meeting consumer demand for power had been considered. . . . I wanted assurances that this power line was a critical last resort, not an optional first choice. . . . I therefore directed the subcommittee staff to conduct an investigation to determine if this power line was actually necessary.
>
> . . . I must state at the outset of this hearing that I am frankly disturbed by the subcommittee staff investigation's tentative findings concerning the motives of the Consumers Power Company in this matter. Let me be very candid. . . . Given the evidence that has been collected by the subcommittee staff, the burden of proof clearly at this point, in my mind, rests with Consumers to convincingly demonstrate that any potential risks associated with this project are balanced by the need to supply its consumers with electricity. . . . There is real question in my

mind about the real agenda of Consumers Power Company. Is the company's purpose really to ensure the reliability of the supply of power to Michigan rate payers? Or is it, rather, to enhance the profitability of its holding company, CMS, and CMS's unregulated subsidiaries?

Wolpe's staff had ascertained that the line in question was not being built in response to Michigan rate payers' needs, but in order to sell and deliver power to an outside area to recoup a loss in the millions that Consumers had incurred when it built a nuclear power plant that was never put into operation.

Besides taking your EMF concerns to the electric utilities and the agencies that control them, approach your local and state public health agencies to let them know you want some EMF safety standards to protect you and your family. A handful of policy makers at the national level, like Wolpe, already support this. At the end of the Michigan hearing, Wolpe established an excellent public health position:

> If there is a demonstrable need for a transmission line, then we may end up having to accept certain risks once we know what those risks are. . . . But, if in fact, there is no need for the facility, no demonstrable public interest to be served, then we should not even have to reach the health question. In fact, I would argue that it's demonstrably violative of the public trust to impose upon people that potential hazard if indeed there is no legitimate public interest to be served.

As we have already stated earlier in the book, the more of a public outcry there is about the hazards of EMFs, the greater the likelihood that our decision makers are going to move on the issue. There are presently no good magnetic field safety standards anywhere in the country and no state has any standards at all for distribution lines. Let your regulators and elected officials know you don't want to be exposed to power-line EMFs over the limits reported by the latest studies — not in your homes, your children's schools, or at work. But you'll have to keep at it to get results. Standard setting is a notoriously slow process.

The same holds true for science. Don't look to the scientific community for a quick fix on this one. But, even if we set aside, for the time being, the desire for scientific consensus, it has been demonstrated that there's ample need for field-effects research to help us get a better grasp on the dangers of EMF exposure. In particular, we need studies of breast cancer and EMFs;

Epilogue

studies of animals that are exposed to EMFs and get cancer; and studies of EMF effects on the known mechanisms of carcinogenesis. California biophysicist Norman N. Goldstein, who teaches one of the only bioelectrophysics courses in the country, says there's a need for research to go back to the sites of the epidemiological studies and determine what frequency and wavelengths actually account for the reported on-site finding. Others in the field have equally interesting ideas for research.

But none of this will happen unless we address the overall question of EMF research funding. There has recently been some improvement as some key studies have been funded: the Stevens study on female breast cancer; a $10 million National Toxicology Program (NTP) EMF investigation that will focus on animal studies; and a large-scale four-year National Cancer Institute (NCI) study on childhood leukemia and EMFs. But, a great deal more funding needs to be made available if we're to establish EMF-effects research as a national priority. For a country that currently spends some $30 billion on cancer treatment, it behooves us to pour a lot more money into a subject that could result in limiting the disease.

It will also have to be the right sort of money. We need to exclude the special interests from the arena and increase federal funding of this research. In the past, EMF studies have been funded, almost exclusively, by the electric industry. And, as Keith Laughlin puts it, "The industry lacks a certain credibility." Others agree. Dr. David O. Carpenter warns, "We need to make sure studies are financed by individuals who do not have any financial interest in their outcomes."

Dr. Robert Becker, a longtime advocate of more federal funding of EMF research, says,

> In an ideal world, what should be done? We need a specific appropriation by Congress, at least $10 million a year, for the next five years. Then we'd have to set up a completely new institute in the NIH (National Institute of Health). You can't have the same people running the granting apparatus who've been testifying on the side of the utilities and the military all these years. Even the National Academy of Sciences, our highest scientific authority, has been compromised. And, most important, the information has to be made public regularly; there must be no holding back of information. There should be some sort of reliable reporting apparatus, something like a yearly contractors' review. We have to open it up. After all, it's tax money, not government money — it's the people's money that's being spent.

At the same time, research and development should continue on means of mitigating fields in power lines, ways to deliver electricity without this rampant EMF exposure, and methods to reduce appliance fields. We also have to find a way to combat the enormous resistance on the part of the utilities to put these methods into use on their lines. The government should require field reductions in appliances. In the words of Herbert M. Kaufman, program manager for ESEERCO, "Certainly, in the future, people who manufacture components that plug into the electric power system should have to make them with low magnetic fields."

Does it seem like an awfully big order? If you're wondering how in the world people like you and me could ever change such a complicated environmental problem, here's a story about a group of people in an Eastern-bloc country that will probably interest you. At a dinner party in San Francisco, early in 1991, a woman read aloud a letter she'd just gotten from her high school pen pal in Romania. The Romanian had written to tell about the people's revolt she'd just taken part in. In the last few days of the revolt, everybody just went into the streets and started to say, "No." The streets were packed with unarmed people, holding hands and shouting, "No! No! No!" while all around them were tanks and soldiers armed with machine guns. But the people didn't care what happened to them; they just had had enough and wanted the dictatorship to end. The men bared their chests to the soldiers, as if to say, "You can shoot us if you want, but we're not going to give up until we get our freedom." If those people could bring down that terrible regime, we can persuade our decision makers to protect us from an environmental health hazard. What it will take is conviction.

Appendix A

Major Studies

The results of an impressive array of studies on the adverse health effects of EMFs have been coming in for more than fifteen years, in countries as diverse as Sweden, Russia, New Zealand, England, and the United States. Reliable researchers have reported consistently troubling findings about the effects of electromagnetic field exposure on tissues, cells, and entire biological systems. There are three types of EMF studies:

- **Epidemiological** — statistical investigations that examine the pattern of a disease in a population or group
- **Whole animal** — in vivo laboratory experiments
- **Cellular laboratory** — in vitro experiments on tissues or cells

EPIDEMIOLOGICAL STUDIES

The earliest studies to warn of the dangers of EMFs were epidemiological. By now, there have been dozens of these studies and more are still going on. These epidemiological studies are grouped according to the population that is being examined:

- Residential EMF exposures and childhood cancer
- Residential EMF exposures and adult cancer
- Occupational EMF environments and cancer

Studies on Children. The major epidemiological studies of children with cancer have consistently found a relationship between EMF exposures at home and three common types of childhood cancer: leukemia, central nervous system cancers (brain tumors), and lymphomas.

WARNING: the electricity around you may be hazardous to your health

- **Wertheimer and Leeper** studied 344 Denver children who died of cancer. This study used wiring code configurations as a surrogate measure of the strength of electromagnetic fields. (Later studies have all reported that wiring code configuration is an acceptable measure of exposure.) Maximum magnetic fields associated with the lines ranged from 4 to 35 mG. Wertheimer found that children exposed to 2 to 2.6 mG magnetic fields were two and a half times as likely to die of leukemia, central nervous system cancers (brain tumors), and lymphomas. The children with cancer were two to three times as likely to have lived near high-current configuration power lines than were the controls. The highest risk was for children in homes with the greatest exposure who had lived in that home for their entire lifetime. (Wertheimer, N. and E. Leeper. "Electrical Wiring Configurations and Childhood Cancer." *American Journal of Epidemiology* 109 [1979]: pp. 273-284.)

- **Wertheimer and Leeper** also studied adult cancer in Denver and found a significant association between some kinds of cancer and proximity to power lines. The association did not hold for leukemia. (Wertheimer, N. and E. Leeper. "Adult Cancer Related to Electrical Wire Near the Home." *International Journal of Epidemiology* 11 [1982]: pp. 345-355.)

- **Tomenius** studied children with cancer in Sweden, using high-voltage (6,000-200,000 V) power lines and high-current electric structures (transmission lines, substations, subways, electric railroads) to determine EMF exposure. He also took measurements of magnetic fields at the front doors of the children's homes. This study found that children living in homes with EMF measurements of 3 mG or above were twice as likely to die of cancer than children whose exposure was lower. (Tomenius, L. "50-Hz Electromagnetic Environment and the Incidence of Childhood Cancers in Stockholm County." *Bioeletromagnetics* 7 [1986]: pp. 191-207.)

- **Savitz,** as part of the New York State Power Lines Project, was charged with replicating the Wertheimer study on a different population of 252 Denver children with cancer and 222 controls. Savitz examined possible confounders and found that none of them, including the magnetic fields from household appliances, influenced the results. He measured fields inside the homes at times of both low- and high-power usage and reported that cancer incidence related solely to ambient fields caused by external power lines, not appliance use. The study confirmed the correlation between wiring codes and actual measurements of magnetic fields: The average field in high-current configuration (HCC) homes was 2-3

Appendix A: Major Studies

mG. Savitz found double the risk for the same site-specific cancers (i.e., cancers that occur at specific anatomical sites) as Wertheimer had reported and double the cancers at all other sites for children who lived near high-current distribution lines. Children who lived in homes with the highest current had a five times greater risk than the controls. Children who had lived in the same house from birth to death had the greatest risk. The author also reported a statistically significant relationship between EMF dosage, based on wire code configurations, and cancer. Savitz had gone on record stating that he believes twenty percent of all the childhood cancer in the United States is due to electromagnetic field exposure. (Savitz, D.A. et al. "Case-Control Study of Childhood Cancer and Exposure to 60-Hz Magnetic Fields." *American Journal of Epidemiology* 128 [1988]: pp. 10-20.)

- **Savitz** reported a 300-percent increase in brain tumors in children whose mothers had used electric blankets during their pregnancies. (Savitz, D. A. et al. "Magnetic Field Exposure from Electrical Appliances and Childhood Cancer." *American Journal of Epidemiology* 131 [1990]: pp. 763-773.)

- **Peters**, in accordance with California Senate Bill 2519, which directed the State Public Service Commission and the State Department of Health Services to study EMFs, added an EMF component to an ongoing University of Southern California NIH-funded study of childhood brain cancer. According to a report by the PUC to the state legislature, "(The Peters study), if positive, would probably be viewed as definitive and would have profound regulatory implications." The study confirmed the Wertheimer and Savitz findings: Children with EMF exposure of 2 to 3 mG in their homes have a two and a half times greater risk for leukemia. The study also found wire code proxies to be a better measure of risk than room measurements. Peters also found that children's use of black-and-white televisions and hair dryers doubled their risk for the disease. When the earth's magnetic field was taken along with the 60-Hz magnetic field measurement, children who were exposed to more than 3-mG fields had six times the risk for cancer. (Peters, J. "Childhood Leukemia and Exposure to Electricity." *American Journal of Epidemiology* [November, 1991]: pp. 215-230.)

- **Peters and Bowman** reported that when the USC 60-Hz data were combined with measurements of the earth's DC magnetic field, researchers had obtained greatly increased risk ratios, based on the 60-Hz fields. In some cases, for the higher AC exposure groups, the risk was from six to

nine times the expected incidence, rather than the two or two-and-a-half increase linked to the AC field alone. (Bowman, J.D. "The Risk of Childhood Leukemia from Home Exposure to Resonance from Static and Power Frequency Magnetic Fields." Abstract presented at DOE conference. November, 1991.)
- **Kraut** (University of Manitoba) did a large epi-study of Canadian children and found as association between leukemia and brain cancer rates and residential electricity consumption. The more electricity used in the household, the higher the rate of cancer. (Kraut. "Archives of Environmental Health," vol. 49, [May, 1991]: pp. 156-159.)
- A large residential study using data from the cancer registry in Taiwan — **Chung-Yi Li, Gilles Theriault, Ruey Lin** — found increases in adult leukemia linked to EMF exposures over 2 mG and distances from power lines. Exposure to fields of 2 mG and higher led to a forty to seventy percent greater risk of leukemia. They also found a significant association between risk and distance from power lines; (same as the 1992 Karlinski childhood study). When cases lived fifty meters or closer to 69-kV lines, they had twice the leukemia risk. (*Epidemiology* 8. [January, 1997]: pp. 25-30.)
- **Lin** studied childhood cancer mortality in Taipei, China, and found elevated risks for the same kind of tumors as in the Wertheimer study in children with magnetic field exposures over 2 mG. Children who lived in the same house from birth to death had the highest risk. The study reported a dose/response ratio, as seen in Figure A. (Lin, R.S. "Childhood Leukemia in Relation to Residential Magnetic Fields." Paper presented at DOE annual review of biological effects of 50- and 60-Hz electromagnetic fields,. 3 November 1991.)
- **Spitz and Johnson, and Wilkins**, in separate studies, examined children with cancer whose fathers worked in electrical occupations and found a link between childhood cancer and the occupation of parents (which suggests, but doesn't prove, possible genetic damage caused by high EMF exposure). Both studies found these children had greater risks of neuroblastoma and brain cancer if their fathers were exposed to high EMFs. Children of fathers with EMF occupational exposure had two and a half times the risk for tumors. For electronic workers' children, the risk was over ten times as high. (Spitz, M. and C. Johnson. "Neuroblastoma and Paternal Occupation." *American Journal of Epidemiology* 121 [1985]: pp. 924-929. Wilkins, J.R. et al. "Paternal Occupation and Brain Cancer of Offspring: A Mortality-Based Case-Control Study." *American Journal of Industrial Medicine* 14 [1988]: pp. 299-318.)

Appendix A: Major Studies

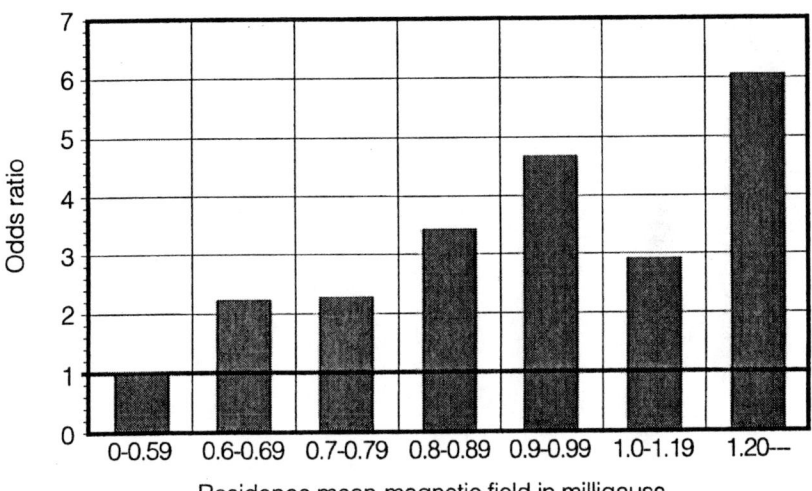

An odds ratio of 1 is the norm.
Graph of data by ELF Magnetic Surveys.

- **Albom and Feychting** reported that exposure to high-voltage power-line magnetic fields increases the risk of cancer. The closer a family lived to the lines and the higher the MFs in their home, the greater the risk (*Environmental News Board*, October, 1992.)

- In 1995, **Albom and Feychting** revisited their study, combining and analyzing their data with that of a 1992 Danish study by Dr. Jorgen Olsen, who had also reported increased cancer risks with EMF exposures. The meta-analysis confirmed the risks for cancer for children with EMF exposures above 2 mG — choosing the cutoff point is important — and found a dose/response link. The higher the field, the greater the risk. The combined analysis reported twice the leukemia risk for children exposed to 1-mG fields, four times the risk for fields of 4 mG, and in fields of 5 mG and over, five times the risk of leukemia, lymphoma, and brain cancers. (*European Journal of Cancer*. [November, 1995].)

- **Feychting, Floderus and Forssen** combined adult subject EMFs at home and at work and found that people with the highest combined EMF exposure had three to four times the risk of leukemia. (*Epidemiology*, vol. 8.[July, 1997]: pp. 384-389.)

- **Roger Cognill**, in a United Kingdom childhood leukemia study, found a causal connection with electric field exposures — not the magnetic fields. (Cognill also believes crib deaths and other neurological and behavioral disorders are linked to EFs.) He took twenty-four-hour measurements in the children's bedrooms. Measured MFs were low, but EFs were almost twice as high as the controls. (*American Journal of Epidemiology*, 144 [2] [1996]: pp. 150-160.)

- Two studies have linked diseases to the EMFs from sewing machines, both at home and at work. A 1991 study by **Dr. Claire Infante-Rivard** at McGill University (Montreal, Canada) found seven times the incidence of leukemia in children whose mothers used sewing machines while pregnant (*Journal of Epidemiology and Community Health*, 45. [1991]: pp. 11-15.)

- A 1994 study by **Dr. Eugene Sobel**, (USC, Los Angeles) linked Alzheimer's disease to EMFs from sewing machines and other sources. (Dosimetry experts have subsequently measured EMFs in the hundreds of mG around sewing machines. (*American Journal of Epidemiology*, 1 September 1995.)

- **Wertheimer and Leeper** found that when pregnant women used electric blankets and waterbeds, fetal growth was affected and they had a higher incidence of miscarriages. The investigators reported that the highest risks occurred in winter, when use of electric blankets was greater. (Wertheimer, N. and E. Leeper. "Possible Effects of Electric Blankets and Heated Waterbeds on Fetal Development." *Bioeletromagnetics* 7 [1986]: pp. 13-22.)

- **Schuz and Michaelis** (University of Mainz) found that children with MF exposures of 2 mG and higher had double the risk of leukemia. The risk was greater for children under four and increased with nighttime exposures. (*Cancer Causes & Control*, vol. 8. [1997]: pp. 167-174.)

OCCUPATIONAL STUDIES

Studies of workers in the electric and electronic industries — electrical engineers, radio and telegraph operators, telephone operators, electricians, power company and telephone linemen, television and radio repairpeople, motion picture projectionists, streetcar and subway motorworkers, power station operators, welders, flamecutters — have reported higher levels of cancer and increased cancer mortality than those occurring in the general population. In particular, increased risks have been reported for leukemia, lymphoma, central nervous system cancers (brain cancer), and malignant melanoma of the skin.

Appendix A: Major Studies

- **Milham** found an increase in leukemia deaths in electrical workers. Electricians, power station operators, and people who worked in the aluminum industry (very high magnetic fields are produced by the aluminum manufacturing process) had the highest risks. (Milham, S. "Mortality from Leukemia in Workers Exposed to Electric and Magnetic Fields." *New England Journal of Medicine* 307 [1982]: p. 249.)

- **Bowman** measured EMFs in the workplaces that Milham had identified and reported higher fields than those in other workplaces or homes. (Bowman, J.D. et al. "Exposures to Extra Low Frequency EMFs in Occupations with Elevated Leukemia Rates. *Applied Industrial Hygiene* 3[6] [1988]: pp. 189-193.)

- **Wright** found increased leukemia risks for workers in Los Angeles with occupational exposure to EMFs, with highest risks of acute myeloid leukemias for power company and telephone linemen. (Wright et al. "Leukemia in Workers Exposed to Electric and Magnetic Fields." *Lancet* 1160 [1982] p. 61.)

- **McDowall** reported excess leukemia risks for electrical engineers, electronic engineers, telegraph operators, and electricians in England. (McDowall, M.N. et al. "Leukemia Mortality in Electrical Workers in England and Wales." *Lancet* 1 [1983]: p. 246.)

- **Howe and Lindsay** studied cancer incidence in various occupations in Canada. They found that transportation and communications workers, who have high EMF exposure, had a higher risk of getting twelve different kinds of cancer. These workers had the highest risk of dying from leukemia and malignant neoplasm. The group with the greatest exposure to EMFs, linemen and repairpeople, had more cases of leukemia, cancer of the stomach, and cancer of the intestine. Workers in the appliance manufacturing industry had the greatest risk of dying from lymphoma or leukemia. (Howe, G.R. and J.P. Lindsay. "Cancer Mortality in Males." *JNCI* 70 [1983]: pp. 37-44.)

- **Lin** found an association between occupational EMF exposures and primary brain cancer (tumors that originate in the brain, rather than travel there from other sites in the body) mortality in adults in Maryland. Electricians and electronic engineers or technicians were at greater risk when cases were grouped according to EMF exposure levels, as ascertained from their job descriptions. Those with the highest exposures had the highest cancer mortality and workers with the highest EMF exposures also died at a younger age. Workers with the highest EMF exposure had twice the expected tumor rate. (Lin, R. et al. "Occupational

Exposure to EMFs and the Occurrence of Brain Tumors." *Journal of Occupational Medicine* 27 [1985]: pp. 1413-1414.)
- **Savitz and Calle** studied leukemia mortality in workers and found significant excess mortality for radio and telegraph operators and electrical engineers. Increased mortality was also reported for linemen, welders, flamecutters, electrical engineers, and electronics technicians. The authors also reviewed the data from eleven leukemia studies and found that electrical equipment assemblers, aluminum workers (routinely exposed to magnetic fields of 100 mG), and telegraph, radio, and radar operators had a higher risk of the disease. (Savitz, D.A. and E.E. Calle. "Leukemia and Occupational Exposure to EMFs: Review of the Epidemiological Surveys. *Journal of Occupational Medicine* 29, no. 1 [January, 1987].)
- **Coleman and Berel** did a similar review and reported that electrical workers had an eighteen percent greater risk for leukemia. (Coleman M. and V. Berel. "Review of Epidemiological Studies of the Health Effects of Living Near or Working with Electricity Generation or Transmission Equipment." *International Journal of Epidemiology* 17 [1988]: pp. 1-13.)
- **Gilman** found excess leukemia mortality in U.S. miners exposed to EMFs from overhead electrical distribution wires in the mines. (Gilman, P.L. "Leukemia Risk Among U.S. White Male Coal Miners." *Journal of Occupational Medicine* 27, no. 9 [1985]: pp. 669-671.)
- **Pearce** found an excess of leukemia in men in New Zealand whose occupations had high EMF exposures. Radio and TV repairpeople and electricians showed the greatest risks. (Pearce, N.E. et al. "Leukemia in Electrical Workers in New Zealand." *Lancet* 1 [April, 1985]: pp. 811-812.)
- **Flodin** found increased risk for leukemia in Swedish electrical workers: electrical technicians, welders, and computer and telephone mechanics. (Flodin. U. et al. "Background Radiation, Electrical Work, and Some Other Exposures Associated with Acute Myeloid Leukemia." *Archives of Environmental Health* 41, no. 2 [1986]: pp. 77-84.)
- **Stern** found an excess of acute myelogenous leukemia in Swedish electrical workers. (Stern, F.B. "A Case-Control Study of Leukemia at a Naval Shipyard." *American Journal of Epidemiology* 123, no. 6 [1986]: pp. 980-992.)
- **Lin** studied electric power industry employees in Taiwan and Maryland and reported increased mortality from liver cancer, brain tumors, and leukemia. The study reported an association between occupational EMF

exposures and primary brain cancer (tumors that originate in the brain, rather than travel there from other sites in the body) mortality in adults in Maryland. Electricians and electronic engineers or technicians were a greater risk when cases were grouped according to EMF exposure levels, as ascertained from their job descriptions. Those with the highest exposures had the highest cancer mortality and workers with the highest EMF exposures also died at a younger age. Workers with the highest EMF exposure had twice the expected tumor rate. (Lin, R. et al. "Occupational Exposure to EMFs and the Occurrence of Brain Tumors." *Journal of Occupational Medicine* 27 [1985]: pp. 1413-1414.)

- **Thomas** compared deaths from brain cancer and central nervous system cancer of electrical workers to the general population and reported elevated risk for workers in the fields of manufacturing and installation and repair of electrical and electronic equipment. The author also found a relationship between incidence of brain cancer and duration of employment in electrical occupations with high EMF exposure. Workers with both RF and ELF exposure had double the risk of brain tumors. (Thomas, T.L. "Brain Tumor Mortality Risk Among Men with Electrical and Electronic Jobs." *JNCI* 79, no. 2 [1987]: pp. 233-238.)

- **Speers** reported that transportation and communication workers had increased risk of brain tumors. Workers in electric power utilities (substations) had thirteen times the incidence. When occupations were grouped according to exposure, there was a significant trend for higher risk with greater exposure. (Speers, M.A. et al. "Occupational Exposures and Brain Cancer Mortality." *American Journal of Industrial Medicine* 13 [1988]: pp. 629-638.)

- **Linet** found that electrical line workers had a significantly greater risk for chronic lymphocytic leukemia. (Linet, M.S. et al. "Leukemia and Occupation in Sweden." *American Journal of Industrial Medicine* 14 [1988]: pp. 319-330.)

- **Matanowski,** in an ongoing study of 50,000 New York telephone company employees with EMF exposure, found increased risk for total cancers, including seven times as much leukemia as in other occupations. Matanowski found twice as many brain tumors among workers with the highest EMF exposures (cable splicers). The authors also reported significant increases in male breast cancer. (Two other studies, including one by Paul Demers of the Fred Hutchison Cancer Research Center in Seattle, Washington, also reported increased risks of male breast cancers linked to EMF exposures.) Matanowski also reported a dose/response ratio — that is, groups with the highest EMF exposures exhibited the highest risk. (Matanowski, G. et al. "Leukemia in Telephone Company Employees." *Contractors Review.* U.S. D.O.E./EPRI. [1988].)

- **Savitz and Loomis** studied occupational mortality data from sixteen states and found excess risks for brain cancer for all electrical workers, with highest risks for electronic technicians and electric power repairers and installers. The authors found excess risks of leukemia mortality for electricians and electrical and electronic technicians. Reviewing mortality rates in 1,000 men, they found that electrical workers died of malignant brain cancer at a fifty percent greater rate than did men in other occupations. (Loomis, D.P. and D.A. Savitz. "Brain Cancer and Leukemia Mortality Among Electrical Workers." *American Journal of Epidemiology* 130 [1989]: p. 814.)

- In 1995, **Savitz and Loomis** expanded their study to 140,000 utility workers and reported a link between EMF on-the-job exposures and brain cancer — but no increased risk of leukemia. They also found forty percent higher breast cancer rates for women in electrical occupations. (Savitz. "Magnetic Field Exposure in Relation to Leukemia and Brain Cancer Mortality Among Electrical Utility Workers." *American Journal of Epidemiology.* [20 January, 1995].)

- In 1997, using the same 1995 data base, **Savitz** reported increases in Lou Gehrig's disease (ALS) and a small, but significant, increased risk of Alzheimer's disease.

- **Theriault** studied electric workers with high on-the-job EMF exposures and reported that the men had increased risks for brain cancer and leukemia. This study involved 223,000 workers for Hydro Quebec, Electricite de France, and Ontario Hydro. (*American Journal of Epidemiology.* [15 March 1994].)

- In 1996, **Dr. Anthony Miller** reexamined the large 1994 Hydro Quebec-Ontario Hydro study and found increased leukemia rates linked to electric field as well as magnetic field exposure. When workers' exposures to both electric and magnetic fields were taken into account, the increased risk for leukemia was eleven times the expected rate — the highest increased risk yet reported by any epi-study. (*American Journal of Epidemiology.* [15 July 1996]: pp. 150-160.)

- In 1996, **Guenel** et al. examined the Electricite de France data and reported an increased risk for brain tumors and colon cancer for workers with electric field — not MF — exposures. This study found no association with leukemia.

Appendix A: Major Studies

- For decades, the Soviets have been reporting on dire health effects from EMF occupational exposures, but until very recently, their studies have been generally disregarded in the United States. In a series of studies of men in electrical occupations, switchyard workers in particular, Soviet researchers have found altered blood pressure, chronic stress effects, immune system dysfunctions, changes in white and red blood cell count, increased metabolism, stimulation of the thyroid, chronic fatigue, and headaches.

- **Juutilainen** studied electrical workers in Finland and reported that the highest risk for leukemia was for linemen and cable joiners. (Juutilainen, J. et al. "Results of an Epidemiological Cancer Study Among Electrical Workers in Finland." *Journal of Bioelectromagnetism* 7, no. 1 [1988]: pp. 119-121.)

- **Nordstrom** found abnormal development in human fetuses of fathers who worked in high-voltage electrical substations. (Nordstrom, et al. "Effects of Paternal EMF Exposure on Offspring." *Bioelectromagnetics* 4 [1983]: pp. 91-101.)

- **Nordenson** reported that workers in a 400-kV electrical substation in Sweden, where they are regularly exposed to short, extremely high magnetic fields and switching surges that cause spark discharges, had elevated rates of chromosomal breaks in their lymphocytes. (Nordenson, I. et al. "Effects in Human Lymphocytes of Power Frequency Electric Fields." Radiat. Environ. Biophys. 23 [1984]: pp. 191-201.)

- **Dr. Patricia Coogan** (Boston University School of Medicine) reported that women who work near mainframe computers and other equipment that generate strong magnetic fields have increased risk of breast cancer. Women whose jobs exposed them to high EMFs had a forty-three percent or one and a half times greater risk of breast cancer. (*Epidemiology*. [September, 1995].)

- **Szmigielski, Vagero, and Tornqvist** found excess skin cancer mortality in jobs with EMF exposures, even in men who never worked outdoors in the sum. (Szmigielski, S. et al. "Immunological and Cancer-Related Effects of Exposure to Low-Level Microwave and RF Fields. Modern Electricity, edited by Marino. New York: Marcel Dekker [1987]. Vagero, D. et al. "Cancer Morbidity Among Workers in the Telecommunications Industry". *British Journal of Industrial Medicine* 42 [1985]: pp. 191-198. Tornqvist, S. et al. "Cancer in the Electric Power Industry." *British Journal of Industrial Medicine* 43 [1986]: pp. 212-213.)

- In 1997, **Dr. Leeka Kheifets and Dr. Patricia Buffler,** et al., conducted a meta-analysis of electrical workers and found increased cancer risks; most significantly a 20% increase in brain cancer. (The studies used in the Kheifets study had reported increased brain tumor risks with EMF exposures.) (*Journal of Occupational and Environmental Medicine.* [December, 1995].)
- **Reif** studied brain cancer in a number of occupations with EMF exposures in New Zealand and reported increased risks for electrical engineers and electricians. (Reif, J.S. et al. "Occupational Risks for Brain Cancer." *Journal of Occupational Medicine* 31, no. 10 [1989]: p. 863.)
- **Mack, Preston-Martin, and Peters** found significantly elevated risks of primary brain tumors (astrocytomas) for men in Los Angeles with occupational exposure to EMFs, with highest risks for electricians and electrical engineers. When brain tumor risk was correlated with duration of employment, a significant upward trend was found after ten years of work in an occupation with high EMF exposures. (An earlier USC study also reported greatly elevated risks of leukemia for men in electrical occupations.) The authors stated, "Our results confirm the findings of a positive association between employment in jobs involving presumed EM-field exposure and brain tumor risk. Greater confidence can be placed in the validity of our results because of the detailed and comprehensive job histories that were obtained directly from the subjects." This study also found a correlation between increased risk for chronic myeloid leukemia and employment as a welder. (Mack, C. et al. "Astrocytoma Risk Related to Job Exposure to Electric and Magnetic Fields." *Bioelectromagnetics* 12 [1991]: pp. 57-66.)
- **Garland** found excess risk of leukemia for U.S. Navy electricians' mates. Reporting in the American Journal of Epidemiology, the authors noted that the finding "should be considered in the context of the literature supporting an association between exposure to electromagnetic fields and increased risk of leukemia." (Garland, E. et al. "Incidence of Leukemia in Occupations with Potential EMF Exposure in U.S. Navy Personnel." *American Journal of Epidemiology* 132, no. 10 [1991]: p. 293.)
- A 1996 Oxford University study (**Fear,** et al) found that British electrical workers had higher rates of brain cancer and leukemia. Using cases from the national cancer registry, the study found increased cancer risks as high as twenty percent. (Fear, et al. "Electrical Workers: An Analysis of Cancer Registration in England." *British Journal of Cancer.* [1996]: pp. 935-939.)

Appendix A: Major Studies

- **Nordenson** reported that workers in a 40-kV electrical substation in Sweden, where they are regularly exposed to short, extremely high magnetic fields and switching surges that cause spark discharges, had elevated rates of chromosomal breaks in their lymphocytes. (Nordenson, I. et al. "Effects in Human Lymphocytes of Power Frequency Electric Fields." Radiat. Environ. *Biophys.* 23 [1984]: pp. 191-201.)

REPRODUCTIVE EFFECTS

- A large study by **Goldhaber** at Kaiser Permanente Hospital in Oakland, California, looked at women who were exposed to EMFs from their computers and found that women who used VDTs twenty hours or more per week had twice the number of miscarriages and increased birth defects in their offspring. (Goldhaber. American Journal of Industrial Medicine 25 [1988]: pp. 150-155.) A later study by Heitanen found a dose/response relationship between VDT fields and miscarriages. (Heitanen, M. *VDT News* [March/April, 1992], p. 1.)

- **Delgado and Leal** exposed fertilized chicken eggs to magnetic fields that approximated those given off by VDTs and reported serious fetal deformities. (This henhouse study was replicated in 1986 in six laboratories around the world. Five of the six laboratories reported similar developments defects in chicks exposed to ELF-pulsed magnetic fields.) (Delgado, J.N. and J. Leal et al. "Embryonic Changes Induced by Weak Extra-Low Frequency EMFs." *Journal of Anatomy* 134 [1982]: pp. 533-551.)

- **Paullson** reported congenital fetal malformations in mice exposed to VDT frequency EMFs. (Paullson, B. "Working with Visual Display Units." International Labor Office, Geneva. [1989].)

- **Wertheimer and Leeper** found that when pregnant women used electric blankets and waterbeds, fetal growth was affected and they had a higher incidence of miscarriages. The investigators reported that the highest risks occurred in winter, when use of electric blankets was greater. (Wertheimer, N. and E. Leeper. "Possible Effects of Electric Blankets and Heated Waterbeds on Fetal Development." *Bioeletromagnetics* 7 [1986]: pp. 13-22.)

- **Nordstrom** found abnormal development in human fetuses of fathers who worked in high-voltage electrical substations. (Nordstrom, et al. "Effects of Paternal EMF Exposure on Offspring." *Bioelectromagnetics* 4 [1983]: pp. 91-101.)

- **Lindbohm and Hietanen** (Institution of Occupational Health, Helsinki) linked miscarriages to VDT use. The higher the MF and the longer the exposure — the greater the risk of miscarriage. *American Journal of Epidemiology.* [14 November 1993].)

IN VIVO STUDIES

Many live animal studies have found EMF exposures linked to:

- Changes in cell growth and key cellular functions
- Increased tumor growth
- Central nervous system effects, including changes in the production of important hormones
- Reproductive disorders: birth defects and increases in miscarriage rates
- Changes in blood chemistry
- Behavioral changes

Tumor Production

- **McLean and Stuchly** treated mice with a known tumor promoter (TPA) and exposed them to 60-Hz fields. They reported a significant increase in tumor growth, but in following experiments, they weren't able to reproduce the effect. (Stuchly. Report in *Microwave News.* Sept./Oct., 1991. p. 3.)
- **Craig Byus** replicated the Stuchly experiment in 1996 and reported a co-promotional effect: a significant increase in tumors in mice who were exposed to 60-Hz MFs and treated with a tumor promoting chemical. (1996 BEMS meeting report.)
- **Barabiroli** reported increases in oncogenic activity in rats' livers after EMF exposures.
- **Leung** exposed rats to fields and treated them with cancer promoters and reported increased mammary tumors. (Leung, F.C. et al. "Evidence of Stress in Rats Exposed to 60-Hz Electric Fields." *U.S.D.O.E. Contractors' Review.* [1988].)
- **Parola** reported EMF exposures caused changes in chicken embryo fibroblast cells. (Parola, A.H. et al. "Time Varying Magnetic Fields Cause Cell Transformations." *Biophysical Journal* 53 [1988]: W-Pos. 26.)

Appendix A: Major Studies

- **Wolfgang Loscher & Merke Mevissen** (Hanover, Germany) reported lower levels of melatonin and significantly higher tumor rates in rats exposed to 60-Hz EMFs. (Loscher & Mevissen. "Animal Studies on the Role of 50/60-Hertz Magnetic Fields in Carcinogenesis." *Life Sciences* v. 54. [1994]: pp. 1531-1543.)
- In 1994, **Holmberg** (Sweden) reported that mice treated with a carcinogenic chemical and exposed to 50-Hz magnetic fields had increased tumors.
- **Lai and Singh** (University of Washington, Seattle) exposed rats to 60-Hz magnetic fields for two hours and found increased single strand DNA breaks in their brains. (*Bioelectromagnetics*, v. 18, no. 2, [1997].)

Reproductive Effects

- **Marino and Becker** reported as early as 1976 on birth deformities, stunted growth, and increased infant mortality in generations of mice and rats exposed to 60-Hz electric fields. (Marino, A.A. and R.O. Becker. "Power Frequency Electric Field Induced Biological Changes in Successive Generations of Mice." *Experimentia* 35 [1980]: pp. 309-311.)
- **Richard Phillips**, at Battelle Pacific Northwest Laboratories in Richland, Washington, has replicated the Marino experiments in a series of three-generational studies with miniature swine, rats, and mice. Phillips reported double the incidence of birth defects in particular exposed generations of swine, rats, and mice, as compared to controls. (Phillips, R.D. "Biological Effects of Electric Fields on Miniature Pigs." Proceedings of the fourth workshop of the U.S./U.S.S.R. scientific exchange on physical factors in the environment. *National Institute of Environmental Health Science*, [21-24 June, 1983].)
- **Martin** reported that exposure to pulsed ELF electromagnetic fields caused a significant increase in abnormalities in developing chicken embryos. (Martin, A.H. "Magnetic Fields and Time Dependent Effects on Development." *Journal of Bioelectromagnetics* 9 [1988]: pp. 393-396.)

Central Nervous System Effects

Many animal studies have reported central nervous system effects, including changes in the production of key central nervous system neurotransmitters. These are hormones that transmit messages from the brain to cells to regulate critical cellular processes, such as learning, motor activity, and the emotions.

- **Hansson,** in separate experiments, exposed rabbits to outdoor magnetic fields at a power substation and in the laboratory. He found the rabbits with exposures outdoors developed severe structural deformities in their brains. He also reported similar changes in the animals with indoor exposures. (Hansson, H.A. "Purkinje Nerve Cell Changes Caused by Electric Fields." *Medical Biology* 59 [1981]: pp. 103-110.)
- **Albert** replicated the Hansson experiments by exposing rats to electromagnetic fields in the laboratory and reported similar changes in the animals' brain cells. (Albert, E.N. et al. "Electron Microscope Observation on Rat Cerebellum and Hippocampus After Exposure to 60-Hz Electric Fields." In abstracts from the sixth annual meeting of Bioelectromagnetism Society.[1994]: D 4-6, 52.)
- In a series of important studies begun nearly twenty years ago at the Brain Research Institute at UCLA, then continued at the Pettis VA Hospital in Loma Linda, California, **Adey** examined the biological effects of ELF (power-line frequency) EMF exposures on animals and cells. In experiments with rats and cats exposed to ELF pulsed frequencies, Adey and Bawin found changes in the firing rate of brain cells in exposed animals, changes in their EEGs, lowered behavior on tests they'd been trained to perform, as well as changes in overall behavior. (Adey, W.F. et al. "Effects of Weak Amplitude Modified Microwave Fields on Calcium Efflux." *Journal of Bioeletromagnetics* 2 [1981]: pp. 295-307.)
- **Marino and Becker,** in a series of experiments throughout the seventies, exposed rats and mice to EMFs and reported that the fields acted as a biological stressor. (Marino, A.A. and R.O. Becker. "The Effect of Continuous Exposure to Low Frequency Electric Fields on Three Generations of Mice." *Experimentia* 32 [1976]: pp. 505-507.)
- **Takashima** reported changes in EEGs of rabbits after limited EMF exposures. (Takashima, S. et al. "Effects of Modulated RF Energy on the EEGs of Mammalian Brains." Rad. Environ. *Biophys.* 16 [1979]: pp. 15-27.)
- **Salzinger** reported permanent changes in learning ability and performance of rats who'd been exposed to 60-Hz EMFs.
- **Lai and Singh** (University of Washington, Seattle) reported impaired learning in rats exposed to MW radiation.
- Similar behavioral effects have been reported in humans. In a study at the Pensacola Naval Laboratories, **Gibson** reported that when he exposed

Appendix A: Major Studies

human subjects to 60-Hz and 45-Hz EMFs, their short-term memory was impaired. (Gibson, R.S. and W.F. Maroney. "The Effect of ELF Magnetic Fields on Human Performance: A Preliminary Study." NAMPL 1105, AD005898, 1974.) Stollery and Graham each found exposure of humans to 50- and 60-Hz fields altered their performance. (Stollery, B.T. "Human Exposure to 50-Hz Electric Currents." In L.E. Anderson et al., "Interaction of Biological Systems with Static and ELF Electric and Magnetic Fields." CONF-841041, pp. 445-454, National Technical Information Service, Springfield, Virginia. [1987]. Graham, C. et al. "A Double-Blind Evaluation of 60-Hz Field Effects on Human Performance, Physiology, and Subjective States.". In L.E. Anderson et al, ibid.: pp. 471-485.)

- **Seegal** exposed monkeys to 60-Hz electric and magnetic fields and found changes in the production of neurochemicals. (Seegal, R.F. "Chronic Exposure of Primates to 60-Hz Electric and Magnetic Fields." *Bioelectromagnetics* 10 [1989]: pp. 289-300.)

- **Ossenkopp** exposed mice to 60-Hz fields at night and found the painkilling effects of morphine on them was reduced. This effect did not occur with daytime exposure, so the researcher hypothesized that it involved the pineal gland. The study also reported a clear dose/response relationship between the strength of the fields and the effect. (Ossenkopp, K.P. et al. "Reduced Nocturnal Morphine Analgesia in Mice Following a Magnetic Disturbance." *Neuroscience Letters* 40 [1983]: pp. 321-325.)

- **Jaffe** exposed rats to 10-V 60-Hz fields and reported increased excitability — the ability to react to electrical stimuli at the cellular level — in the neurons in their brains. (Jaffe, L.F. and M. Poo. "Neurites Grow Faster Toward the Cathode Than the Anode in a Steady Field." *Journal of Experimental Zoology* 209 [1979]: pp. 115-128.)

- **Phillips** reported the suppression of melatonin in rats exposed to 60-Hz fields. Melatonin is a key hormone produced by the pineal gland. (Phillips, R.D., L. Anderson, and W.F. Kaune. "Biological Effects of High-Strength Electric Fields on Small Laboratory Animals." DOE/RL01830/T. [1987].)

- **Welker** reported that small nighttime perturbations or disturbances of static magnetic fields within the level of ambient 60-Hz fields caused significant decreases in melatonin production in rats. "Welker, H.A. et al. "Effects of Artificial Magnetic Fields on the Rat Pineal Gland." *Experimental Brain Research* 50 [1983]: pp. 426-432.)

- **Wilson** reported significant lowering of the neurotransmitters' melatonin and serotonin in rats that were exposed to EMFs. In a separate study, Wilson found decreased melatonin production in people who slept under electric blankets. "Wilson, B.W. et al. "Chronic Exposure to 60-Hz Electric Field Effects in Pineal Functioning in Rats." *Bioelectromagnetics* 2 [1981]: pp. 371-380.)

- **Vasquez** reported changes in neurotransmitter secretions in the brains of chicks and rats after 60-Hz EMF exposures. (Vasquez, B.J., L.E. Anderson, C.I. Lowery, and W.R. Adey. "Diurnal Patterns in Brain Biogenic Amines of Rats Exposed to 60-Hz Electric Fields." *Bioelectromagnetism* 9 [1988]: pp. 229-236.)

- **Wolpaw** exposed primates to EMFs and found significant declines in the concentration of two important neurotransmitters. One of these remained depressed twenty-one days after exposure. (Wolpaw, J.R. et al. "Chronic Effects of 60-Hz Electric and Magnetic Fields on Primate Central Nervous System Function." *New York State Power Lines Project Final Report*. Albany, New York. 1987.

Effects of Cell Division

- **Barnothy, Mastryvkova, and Dyshlovoi each** reported changes in the cells' ability to synthesize DNA, the protein responsible for transmitting hereditary traits. These studies also reported changes in other critical cellular functions in mice exposed to EMFs. (Barnothy, M.F. et al. "Abnormalities in Organs of Mice Induced by Magnetic Fields." Nature 221 [1969]: pp. 270-271. Mastryvkova, V.M. et al. "Effect of a Strong Magnetostatic Field on Proliferation of Duodenal Epithelial Cells in Mice." Biol. Bull. Aca. Sci. USSR 5, no. 3 [1978]: pp. 371-374. Dyshlovoi, V.D. et al. "Effect of EMF on Growth Pattern and Mitotic Activity of Cultured Human Fibroblastoid Cells. Cytol. Gener. 15, no 3 [1981]: pp. 6-9.)

- **Yao** reported chromosome aberrations in corneal cells of Chinese hamsters exposed to microwaves. (Yao, K.T.S. "Microwave Radiation Induced Chromosome Aberrations in Corneal Epithelium of Chinese Hamsters." *Journal of Heredity* 69 [1978]: pp. 409-412.)

- **Nordstrom** did a study of 542 electrical workers in Sweden and reported chromosome breaks in a group of substation workers. (Nordstrom, S. et al. "Reproductive Hazards Among Workers at High-Voltage Substations." Bioelectromagnetics 4 [1983]: pp. 91-101.)

Appendix A: Major Studies

Effects on the Blood

- **Frey** found that pulsed microwave exposures caused changes in the blood-brain barriers of rats. (The blood-brain barrier is the central nervous system mechanism that prevents toxins from entering brain cells.) (Frey, A. H. et al. "Natural Function and Behavior." *Annals of New York Academy of Science* 247[1975]: pp. 433-438.)
- **Beischer** found low-frequency exposure caused an increase in blood triglycerides in humans. (Beischer, D.E. et al. "Exposure of Man to Magnetic Fields Alternating at Extremely Low Frequencies." U.S. Navy Report No. NAMRL-1180. Pensacola, Florida, Naval Aerospace Medical Research Laboratory. 1973.)
- **Prausnitz and Susskind** exposed mice to RF radiation from an air-force radar transmitter and reported that a third of the mice developed cancer of the white blood cells (leukemia). (Prausnitz, S. and C. Susskind. "Effects of Chronic Microwave Irradiation on Mice." IRE Transcripts on Biomedical Electronics 9 [1962]: pp. 104-108.)

IN VITRO STUDIES

Prompted by the findings of epidemiological studies that showed EMF exposure affects the vital functions of the cells of animals and humans, laboratory scientists began to study known mechanisms involved in carcinogenesis. To explain the statistical increases in cancer, epidemiologists have linked to ELF electromagnetic fields, an impressive body of literature describing the effects of EMF exposures on cells has emerged. In these experiments, scientists have looked at the effects of electromagnetic fields on organs, cells, the cell membrane, the central nervous system, the immune system, and the various glands that regulate the production of hormones.

THE CENTRAL NERVOUS SYSTEM

The central nervous system is a logical place to look for electromagnetic field bioeffects, because its functions are controlled by electrochemical phenomena — interactions of electrical charges and chemical neuro-transmitters (hormones that carry messages from the brain to the cells).

Hormones are produced in the body by the various endocrine glands, such as the thyroid and the pituitary. Specific hormones regulate the major

cellular functions, including growth and development, blood composition, the immune system, and a group of responses known collectively as the stress response. Some key central nervous system (CNS) hormones are:
- Serotonin: controls muscles, including the heart and the vascular system.
- Melatonin: regulates circadian rhythms (24-hour cycles of biological activity often referred to as our "biological clock"), depression, pain, and functions of the immune system. Melatonin plays a critical role in many key cell processes, including tumor inhibition and cancer control. Magnetic fields are known to inhibit the body's melatonin production and block the hormone's ability to fight tumors. The so-called melatonin hypothesis is the theory that EMFs cause increases in cancer because they lower the body's production of melatonin. (For more on this, see Chapter 4.)
- Parathyroid: controls cell growth or proliferation.

The following research has reported that electromagnetic fields affect the production and activity of the neurotransmitters:
- **Luben** reported reduced cell response to PTH in mouse bone cells exposed to pulsed magnetic fields of various frequencies. He also reported changes in blood plasma membrane functions. (Luben, R.A. et al. "Effects of Electromagnetic Stimuli Inhibition of Responses to PTH by Low-Energy Low-Frequency Fields." Proceedings of the National Academy of Sciences 79 [1982]: pp. 4180-4184.)

The effects of EMF exposure on enzymes (other chemicals with key CNS functions) have also been studied. These chemicals include:
- Calcium: essential to the regulation of the majority of cellular processes that are controlled by the brain, including the release of neurotransmitters and mitosis (cell division).
- Protein kinases: enzymes involved in the regulation of cell division and other cellular functions.
  - ODC (ornithine decargoxulase): controls cell growth and proliferation. Increases in ODC lead to tumor production, and ODC levels are a marker for cancer.

The following studies reported changes in these chemicals in cells exposed to EMFs:
- **Adey and Bawin** found that chick brains exposed to ELF fields had changes in the release of calcium. They also reported, in other experiments, reductions in protein kinase in human lymphocyte cells, and

Appendix A: Major Studies

increased ODC production in cultured hamster cells and human melanoma cells. They also reported decreased calcium efflux — a calcium flow from inside the cell — after exposure. (Bawin, S.M.. and W.R. Adey. "Sensitivity of Calcium Binding in Cerebral Tissue to Weak Environmental Electric Fields." *Cellular Biology* 73 [1976]: pp. 1999-2003.)

- **Blackman** reported increased calcium flow in chick brains (at particular windows or frequencies) in transmission-line exposure chambers where chicks had been exposed in vivo. He reported that the changes were caused by the magnetic field component, in interaction with the earth's natural static magnetic field. (Blackman, C.F. "A Role for the Magnetic Field in the Radiation-Induced Efflux of Calcium Ions from Brain Tissue in Vitro." *Bioelectromagnetics* 6 [1982]: pp. 327-332.)

- **Dutta** exposed human nervous system tumor cells to electromagnetic fields and found significant increases in calcium efflux (the flow of calcium out of the cells). (Dutta, S.K. et al. "Radiation-Induced Calcium Ion Efflux from Human Neuroblastoma Cells in Culture." *Bioelectromagnetics* 5 [1984]: pp. 71-78.)

- **Byus** found temporarily increased ODC activity in human lymphoma cells exposed to 60-Hz fields. (Byus, C.V. et al. "The Effects of Low-Energy 60-Hz EMF Upon the Growth Related Enzyme ODC." *Carcinogenesis* 8 [1987]: pp. 1385-1388.)

- **Cain** reported double the amount of ODC activity in normal human fibroblasts exposed to 60-Hz EMFs. (Cain, C., R. Jones, and R. Adey. "Agar-Bridge System for Exposing Cell Cultures to Electric Fields." BEMS abstract. [1986]: p. 60.)

- **Aarholt** found the rate of protein syntheses in general was dependent on magnetic-field exposure. (Aarholt, E. et al. "Magnetic Fields Effect Lac Operon System." *Phys. Med. Biol.* 27, no. 4 [1982]: pp. 606-610.)

- **Phillips and Winters** reported increases in transferrin receptors in cells that were exposed to fields. (Phillips, J.L. and W.D. Winters. "Transferrin Binding to Two Human Colon Carcinoma Cell Lines." *Cancer Research* 46 [1986]: pp. 239-244.)

- **Stevens** reported suppressed melatonin production linked to EMF exposure, resulting from changes in circadian rhythms when the exposures were at night. He hypothesized that electric power may increase breast cancer, which is known to be inhibited by melatonin. (Stevens, R.G. "Electric Power Use and Breast "Cancer: A Hypothesis." *American Journal of Epidemiology* 125 [1987]: pp. 556-561.)

- **Liburdy** (Lawrence Berkeley Labs, California) reported that 60-Hz magnetic field exposures prevented the growth-suppressing action of melatonin on human brain cancer cells. (1996 BEMS meeting.)
- **Blackman** also reported reversal of the inhibiting effect of melatonin after exposure to 60-Hz magnetic fields.
- **Drs. John Reif and James Burch** found EMF exposure significantly lowered melatonin levels. (DOE Annual Review. November, 1997.)
- **Russell Reiter** (University of Texas) has written a book on melatonin and was one of the first people to study the hormone. In a number of animal studies Reiter found a link between magnetic field exposures and lowered melatonin levels in rats. Reiter believes melatonin is the key to EMFs and cancer.
- **Arnetz and Berg** (Karlinski Institute) found decreases in melatonin levels with VDT use. *Journal of Occupational and Environmental Medicine.* November, 1996.)
- At the University of Southern California, **Dr. Stephanie London** is conducting a large epi-study of breast cancer and melatonin.

Cell Growth and Differentiation

Mitosis, or cell division, is the biological mechanism by which new cells are formed in the normal processes of growth and repair in our bodies. Mitosis is also involved in malignant cell growth, which causes tumors.

Studies of whole animals are looking at the role EMF exposure has on cancer production, and few have reported so far. For years, laboratory scientists have tried to bridge this gap by focusing on studies of cellular functions that shed some light on how EMFs affect known mechanisms of cancer or tumor production. (See also tumor promotion section in the in vivo studies.)

- **Cain** treated tumor cells with TPA, a known cancer promoter, then exposed them to ELF fields and reported that exposed cells grew larger than cells that were treated, but weren't exposed. (Cain, C., W.J. Thomas, and W.R. Adey. "60-Hz Magnetic Field Effects of C3H10T1/2 Fibroblasts." BEMS abstract. [1990]: p. 19.)

Most of these researchers are trying to find out if low-frequency EMFs induce changes in the genes and chromosomes themselves. In particular, they're searching for evidence that EMF exposure causes chromosomal breaks or aberrations like the ones found when ionizing radiation (like x-rays) causes cancer. To date, most of these studies have failed to find chromosome breaks in cells exposed to ELF electromagnetic fields, with certain exceptions:

- **Adam** reported chromosomal breaks in cultured hamster cells after magnetic field exposure.
- **Eberle and May** exposed Chinese hamster bone marrow cells to static magnetic fields and found increases in sister chromosome exchange (SCE) — an accepted scientific test for chromosome breaks. (Eberle, P. and C. May. "SCE-Frequencies Following Exposure to Magnetic Fields." Workshop Mutations Forschung [1982]: p. 17.)
- **Mastryvkova** reported SCEs in exposed mice cells. (Mastryvkova, V. M. et al. "Effect of a Strong Magnetiostatic Field on Proliferation of Duodenal Epithelial Cells in Mice." Biology Bulletin, Academy of Sciences, USSR 5, no. 3 [1978]: pp. 371-374.)
- **El Nahas** reported changes in mice cells. (El Nahas. "Micronuclei Formation in Somatic Cells of Mice Exposed to 50-Hz Electric Fields." *Environ. Mol. Muta.* 13 [1989]: pp. 107-111.)
- **Tsoneva** found chromosome aberrations in human lymphocytes that had been exposed in vivo. (Tsoneva, M.G. et al. "Effects of Magnetic Fields on Chromosome Set and Cell Division." *Sov. Genetices* 11 [1985]: pp. 398-401.)
- A number of plant studies report chromosome aberrations after extremely low-frequency exposures. (Dubrov, A.P. et al. "Genetic Effect of Static Electric Fields." Dokl. Biol. Sci. 178 [1968]: pp. 29-30. Goswami, H.K. "Changes in Chromosome Morphology Due to Magnetism." *Cytologia* 42 [1977]: pp. 639-644. Herich, R. "Influence of a Magnetic Field on Mitosis and the Structure of the Chromosome." F.R.N. Univ. Com. Physiologia Plantarum 11 [1976]: pp. 1-5.)

Other studies have reported important changes in mitosis or interference with normal DNA and RNA activity (DNA and RNA are the protein molecules involved in cell division):

- **Goodman and Henderson** found changes in RNA transcription and altered protein syntheses in gnat saliva glands they exposed to pulsed magnetic fields. In an ongoing series of cellular experiments at Columbia University and Hunter College, the researchers report enhanced RNA transcription and increased RNA synthesis in cells exposed to EMFs. They also are finding EMF exposures stimulate the genes and alter calcium production. (Goodman, R. and A. Henderson. "Pulsing EMFs

Induce Cellular Transcription." *Science* 220 [1983]: pp. 1283-1385. Goodman, R. and A. Henderson. "Transcriptional Patterns in X Chromosome of Sciara Following Exposure to Magnetic Fields." *Bioelectromagnetics* 8 [1987]: pp. 1-7.)

- **Chang** exposed mouse leukemia cells to static and pulsed microwaves and reported inhibited DNA synthesis. (Chang, B.K.. et al. "Inhibition of DNA Synthesis by Microwave Radiation in L1210 Leukemia Cells." *Cancer Research* 40 [1980]: pp. 1002-1005.)

- **Dyshlovoi** exposed human embryo cells to 50-Hz fields and found that exposure significantly reduced the rate of cell division. (Dyshlovoi, V.D. et al. "Effect of Industrial Frequency EMF on Growth Pattern and Mitotic Activity of Cultured Human Fibroblast Cells." Cytol. Gener. 15, no. 3 [1981]: pp. 6-9.)

- **Liboff** found that DNA synthesis in human fibroblast cells and mouse leukemia cells increased when exposed to ELF fields. (Liboff, A.P. et al. "Time-Varying Magnetic Fields: Effect on DNA Synthesis." *Science* 223 [1984]: pp. 818-820.)

- **Smith** reported that pulsed EMFs could affect tumor growth in mice. The effect depended on the time of application and the distribution of exposure and varied with the sex of the animal. (Smith, S. "Effect of Pulsed EMF on Tumor Growth in Mice." *Journal of Bioelectricmagnetism*, vol. 4, no. 1. [1985].)

- **Phillips and Winters** have conducted a series of experiments exposing human colon and brain cancer cells to 60-Hz fields, and reported that the cancer cells greatly increased. They also found they became more resistant to immune system attack. (Phillips, J.L. and W.D. Winters. "Transferrin Binding to Two Human Colon Carcinoma Cell Lines: Characterization and Effect of 60-Hz EMFs." *Cancer Research* 46 [1986]: pp. 239-244.)

- **Phillips and Winters** reported that 60-Hz field exposures resulted in increased cancer cell proliferation, increased transferrin receptors on cells, and increased production of tumor-associated antigens (enzymes capable of generating an immune response; an increase in them is a measure of tumor growth). They also found that human colon cancer cells greatly increased the rate at which they synthesized DNA after twenty-four hours in 60-Hz magnetic fields. In a separate study, Winters reported 60-Hz fields at power frequency levels accelerated malignant growth of human cancer cells. In one week they were cloning six times as fast. (Phillips, J.L. and W.D. Winters. "In Vitro Exposure to EMFs: Changes in Tumor Cell Properties." *International Journal of Radiation Biology* 49 [1986]: pp. 463-469.)

THE IMMUNE SYSTEM

There have also been studies examining direct EMF effects on the immune system itself. The lymphocytes are the killer cells of the immune system. They defend the body against biological agents of disease, chemical invaders, and cancerous growths. It is important to note that a healthy immune system can provide us with immunity from cancer, just as it does for other diseases. Our lymphocytes have the ability to stop tumor cells from reproducing or to kill them. A number of studies have reported immunosuppressant (preventing the lymphocytes from doing their job) effects of EMF exposures:

- **Lyle** reported a decrease in the ability of mouse T-lymphocytes to attack cancer cells after exposure to 60-Hz fields. The stronger the field, the more it inhibited the cytotoxicity (the ability to fight tumors) of the lymphocytes. (Lyle, V.V. et al. "Suppression of T-Lymphocyte Cytotoxicity Following Exposure to 60-Hz Sinosoidal Electric Fields." Bioelectromagetism 9 [1986]: pp. 303-307.)

RF/MW STUDIES

Radio Frequency (RF) radiation is high frequency nonionizing radiation; Microwaves (MW) are at the highest end of the RF frequency. Many studies have reported dangerous effects from RF/MW exposures. (See Chapter 8.)

- **Milham** found higher leukemia mortality in ham radio operators, as well as greatly increased incidence of other kinds of cancer, including lymphomas and Hodgkin's disease. (Milham. S. "Silent Keys: Leukemia Mortality in Amateur Radio Operators." *Lancet* 1 [6 April 1985]: p. 812.
- **Szmigielski** studied cancer mortality in Polish military officers with high radio frequency (RF/MW) and ELF (power-line frequency) radiation exposures and found death rates that were six times as high as those without exposures. In a different study, Szmigielski related chronic microwave exposure to high blood pressure, headaches, memory loss, and brain damage. (Szmigielski, S. et al. "Immunological and Cancer-Related Effects of Exposure to Low-Level Microwave and RF Fields." In *Modern Electricity*, edited by Marino. New York: Marcel Dekker. 1987.)

- **Henderson and Anderson** in a large case/control study conducted by the Hawaii Department of Health, examined residential exposures of census tracts with (and without) broadcast towers in Honolulu. The study found significantly elevated cancer rates in males living in eight of the nine tracts with broadcast towers, and some increase in female cancers as well. (Broadcast towers emit microwave radiation.) (Henderson, A. et al. "Report: Effects of Broadcast Towers on Residential Cancer Rates." Environmental Epidemiology Program, State of Hawaii Department of Health. 1986.)

- **Lester and Moore** studied the cancer incidence in Wichita and reported it was higher in neighborhoods exposed to the radar beams from two airports. The highest cancer rates were found in areas exposed to both beams. When the authors studied cancer mortality statistics around the United States, they found that counties with air force bases — that is, with exposure to radar beams — had a significantly higher incidence of cancer deaths. (Lester, J.R. and D.F. Moore. "Airport Radar and Incidence of Cancer in Wichita." *Journal of Bioelectricity* 1, no 1. [1984].)

- **Hocking** (Melbourne) reported in 1995 that children living near TV towers (antennas) in Australia had elevated rates of leukemia. (Dr. Hocking was the medical director of the state-owned telephone company Telstra.

- **Thomas** conducted an extensive study of American workers exposed to RF and microwave radiation in San Francisco and reported a significant increase in brain tumors. Thomas found a dose/response rate: the longer the exposure, the higher the risk. ("Brain Tumor Mortality Risk Among Men with Electrical and Electronic Jobs: A Case/Control Study." *NCI Journal*, vol. 79, no. 2. [August, 1987].)

- **Szmigielski** reported microwave exposure caused an increase in tumors in mice treated with a known tumor promoter. (Szmigielski, S. et al. "Accelerated Development of Skin Cancer in Mice Exposed to Microwave Radiation." *Journal of Bioelectromagnetics* 3 [1987]: pp. 179-191.)

- Under the auspices of the University of Washington's long-term rat study, **Guy** exposed generations of rats to pulsed microwaves in frequencies that simulate the highest level of human exposures allowed by current microwave standards. Among other findings, they reported induction of adrenal medulla tumors, significantly increased malignant endocrine and ectocrine tumors, increases in carcinomas and sarcomas. (Guy, A.W. et al. "Effects of Long-Term Low-Level RF Radiation Exposure on Rats." vol. 9, University of Washington. USAFSAM-TR-85. [August, 1985].)

Appendix A: Major Studies

- **Prausnitz and Susskind** exposed mice to RF radiation from an air-force radar transmitter and reported that a third of the mice developed cancer of the white blood cells (leukemia). (Prausnitz, S. and C. Susskind. "Effects of Chronic Microwave Irradiation on Mice." IRE Transcripts on Biomedical Electronics 9 [1962]: pp. 104-108.)

- **Soma-Sakar** (Delhi Institute of Nuclear Medicine and Allied Sciences) exposed mice to microwaves — at 1mw/cm2, a power density that has repeatedly been set as the safe limit for general public exposure — and reported mutagenic effects — changes in DNA in brains and testes. (Soma-Sakar. "Effect of Low Power Microwave on the Mouse Genome: a Direct DNA Analysis. Institute of Nuclear Medicine and Allied Sciences. Delhi, India. [6 August 1992].)

- **Takashima** reported changes in EEGs of rabbits after limited EMF exposures. (Takashima, S. et al. "Effects of Modulated RF Energy on the EEGs of Mammalian Brains." Rad. Environ. Biophys. 16 [1979]: pp. 15-27.)

- **Stodolnik-Baranska** exposed human lymphocytes to pulsed microwaves and reported a number of abnormalities in them that were directly related to the length of the exposures. (Pulsed means that a field is switched on briefly, then turned off, in a regular sequence. RF and microwave radiation is often pulsed. Power-line frequencies are not.) (Stodolnik-Baranska, W. et al. "Effects of Microwaves on Human Lymphocyte Cultures." Proceedings of the International Symposium, Warsaw. [1973]: pp. 189-195.)

- In 1990 **Cleary** (Virginia Commonwealth University) reported proliferation in human brain tumor cells that were exposed for two hours to microwaves. Even after the exposure, the tumor cells continued to grow. (*Bioelectromagnetics* vol. 3. [1982]: pp. 453-466.)

A recent spate of studies examined microwave radiation similar to the signals and emissions produced by the new cellular and wireless phone technologies and reported disturbing findings.

- **Repacholi** (Adelaide, Australia) in a study sponsored by the state telephone company, Telstra, exposed mice to MW radiation similar to the signals from GSM mobile phones (a 900-mHz pulse radiated RF signal). Repacholi reported a significant increase — double the rate — in lymphoma in the exposed mice. This is the first animal study reporting non-thermal MW effects. Dr. Repacholi is head of the WHO EMF Projects. ("Lymphomas in Transgenic Mice Exposed to Pulsed 900-mHz EMFs." *Radiation Research*. [May, 1997].)

WARNING: the electricity around you may be hazardous to your health

- **Maes** et al. reported increased tumor production in human lymphocytes (white blood cells) that were exposed for two hours to the field from a GSM antenna, then treated with a carcinogenic chemical. (The general public wouldn't be likely to experience this exposure, but many workers could.) (*Environmental Molecular Mutagen*, vol. 28. [1996]: p. 26.)
- **Hocking** (Australia) studied cell phone users and found they experienced headaches, especially those using cellular phones. (Hocking, the former medical director for Telstra, advised that children be prevented from using the phones.)
- **Carillo** (Mt. Sinai, Miami) reported pacemaker interference from cellular phones.
- **Hayes** also reported that cell phones — in particular, digital models — caused significant electromagnetic interference with pacemakers. Dr. Hayes advised pacemaker users to keep the instruments away from their chests. (Both studies were funded by the WRT, the cellular industry's research arm.) (*New England Journal of Medicine*, 336. [22 May 1997]: pp. 1473-1479.)

Appendix B

Appliances: What's Safe, What's Not

MAGNETIC FIELDS IN MILLIGAUSS

Appliance	At 4 Inches	At 1 foot	"The Magic Number" At 3 Feet
Clothes dryers	4.8 to 110	1.5 to 29	0.1 to 1
Clothes washers	2.3 to 3	0.8 to 3.0	0.2 to 0.48
Coffee makers	6 to 29	0.9 to 1.2	<0.1
Toasters	10 to 60	0.6 to 7.0	<0.1 to 0.11
Crock pots	8 to 23	0.8 to 1.3	<0.1
Irons	12 to 45	1.2 to 3.1	0.1 to 0.2
Can openers	1300 to 4000	31 to 280	0.5 to 7.0
Mixers	58 to 1400	5 to 100	0.15 to 2.0
Blenders	50 to 220	5.2 to 17	0.3 to 1.1
Vacuum cleaners	230 to 1300	20 to 180	1.2 to 18
Portable heaters	11 to 280	1.5 to 40	0.1 to 2.5
Faust blowers	3 to 120	.25 to 370	<0.1 to 3.1
Hair dryers	3 to 1400	<0.1 to 70	<0.1 to 2.8
Electric shavers	14 to 1600	0.8 to 90	<0.1 to 3.3
Televisions	4.8 to 100	0.4 to 20	<0.1 to 1.5
Fluorescent fixtures	40 to 123	2 to 32	<0.1 to 2.8
Fluorescent desk lamps	100 to 200	6 to 20	0.2 to 2.1
Saber and cirsular saws	200 to 2100	9 to 210	0.2 to 10
Drills	350 to 500	22 to 31	0.8 to 2.0

Source: **Gauger, Jr. Household Appliance Magnetic Field Survey IEEE transactions on power apparatus and systems PA-104 (Stpe. 1985)**

Appendix C

EMF Resources

(Inclusion does not constitute a reommendation or warranty.)

TESTERS

ELF Magnetic Surveys
HCR2, Box 850-295
Tucson, AZ 85735
(800) 749-9873
Email: kriley3@ix.netcom.com

Safe Environments
2512 Ninth Street
Berkeley, CA 94710
(510) 549-9693

National EMF Testing Assn.
628-B Liberty Place
Evanston, IL 60201
(708) 475-3696

AlphaLab
1272 Alameda Avenue
Salt Lake City, UT 84102
(801) 532-6604

CITIZEN GROUPS

Citizens Concerned About EMFs
POB 120
San Ramon, CA 94583

Michigan Safe Energy Fund
Karen Boyer
9944 "O" Drive, South
Burlington, MI 49029

Citizens Against Overhead Power Lines
POB 66045
Seattle, WA 98166

Residents Against Giant Electric (RAGE)
Debbie Studd
4 Hillside
Middletown, NJ 07728

Appendix C: EMF Resources

Health Environments
Herbert Landegger
361 Vernon Avenue, Unit 7
Venice, CA 90291
(310) 452-5354

Environmental Electrics
1618 Grand Avenue
San Rafael, CA 94901
(415) 389-9696

Sage Associates
1283 Coast Village Circle, Ste 5
Montecito, CA 93108
(805) 969-0557

Mary Levenstein
4570 Province Line Road
Princeton, NJ 08540
(609) 683-0692

CA Dept. of Health Services EMF Program
2151 Berkeley Way, Annex 10
Berkeley, CA 94714
(510) 450-3818

Neighbors Opposing Power Encroachment (NOPE)
3846 Magrath Road
Encinitas, CA 92023

Citizens Opposed to Power Exposure (COPE)
2300 Cobleigh Road
Eagle Point, OR 97524

EMR Alliance
410 West 53rd Street
New York, NY 10019
(212) 554-4073

C.A.U.S.E.
POB 493
Antioch, IL 60002

FOR AN AFFORDABLE GAUSSMETER*

Magnetic Sciences, Int'l.
HCR2, Box 850-295
Tucson, AZ 85735
(800) 749-9873
E-mail: kriley3@ix.netcom.com

Ellen Sugarman uses an MSI-25A, which lists for $195.

Glossary

alternating current (AC). An electrical current that changes strength and direction of flow with a regular cycle or alternates. For example, 60 AC is an electrical current that changes its direction and strength sixty times per second.

ampere (AMP). A unit used to measure electrical current.

analog. Using physical variables to represent numbers.

antenna. A device the surface of which is used to capture an incoming and/or to transmit an outgoing RF signal.

cancer cluster. An unusually high number of cancer cases in an area.

carcinogen. A substance that is known to cause cancer.

charge. The electrical property of matter that's responsible for creating electric fields. Charge is either negative or positive.

circadian rhythm The biological cycle that usually occurs at twenty-four-hour intervals and is also known as the biological clock.

circuit. The arrangement of the wires in power lines, wall wiring, and appliances.

conductor. A material that will carry electric current.

confounder. A factor extraneous to a study that distorts the true relationship of the variables being examined.

current. The flow of electric charge through a power line or an electric wire. The current in a line is like the water flowing through a pipe. Currents produce magnetic fields.

Glossary

digital. A clock or watch that shows time by displayed digits or numerals from one to nine.

direct current (DC). An electrical current that does not change strength or direction over time, but remains steady. The current in batteries is DC; so is the earth's natural magnetic field. DC fields are also called static fields.

distribution line. A power line used to distribute electricity locally to a utility's end users.

DNA. Deoxyribonucleic acid. The molecules from which genetic material — genes and chromosomes — is made.

electric field (EF). The forces electric charges exert on other objects. The radiation from an electric source that is produced by voltage.

electromagnetic field (EMF). The field around an electric force that contains an electric field and a magnetic field.

electromagnetic spectrum. A breakdown of electromagnetic fields according to their frequency and wavelength. The spectrum is divided into extra low-frequency (ELF) radiation, very low-frequency (VLF) radiation, radio-frequency (RF) radiation which includes and microwave radiation (MW); light, ionizing radiation.

ELF radiation. Extremely low-frequency radiation at the end of the electromagnetic spectrum from zero to 1,000. The 60-Hz power frequency is in this ELF range.

epidemiology. The study of the distribution of disease in populations.

field. The radiation emanating from an electric source. This may also be called an electromagnetic field.

gauss (G). A unit used for measuring magnetic fields. Since it is such a large measure, the milligauss (mG) is more commonly used. One gauss equals 1,000 milligauss.

gene. The part of the DNA that contains specific hereditary traits.

hertz (Hz). A unit used to measure the frequency of an AC electric current. One hertz is one cycle per second. A 60-Hz system has sixty cycles per second.

WARNING: the electricity around you may be hazardous to your health

hormone. A chemical produced by the glands in the body that controls cellular functions and transmits messages from the brain to the cells.

initiator. A cancer-causing agent such as ionizing radiation or certain chemical carcinogens that cause genetic mutations that can lead to cancer.

kilovolt (kV). One thousand volts. A measure of electric potential.

leukemia. Cancer of the blood tissue.

magnetic field (MF). The force a moving electric charge exerts on other moving charges. The field produced by an electric current.

melatonin. A hormone produced by the pineal gland that controls a number of bodily processes, such as the circadian rhythm, and plays an important role in tumor inhibition.

meta-analysis. The process of combining data from a number of studies and analyzing it statistically.

microwaves (MW). Electromagnetic radiation with a frequency at the high end of radio-frequency radiation. A form of nonionizing radiation.

mitosis. The process of cell division or reproduction.

phase. The timing of an AC electric current. Another name for a hot or energized electric wire. Our 60-Hz current is a three-phase system, with each of the three phases reaching their peak at different points in the cycle. The three work together in transmission lines, distribution lines, and heavy equipment.

pineal gland. A gland, located in the center of the head, that produces melatonin.

promoter. An agent that accelerates the growth of malignant tumors.

radiation. Energy that is propagated through space in waves or particles. Some common forms of radiation are x-rays, microwaves, light, and radio waves.

RNA. Ribonucleic acid. A chemical molecule produced by DNA to transmit its instructions to the cells during mitosis.

Glossary

tesla (t). A very large unit of measure of magnetic fields. One tesla equals 10,000 gauss. A microtesla (µT) — one millionth of a tesla — is a smaller measure that is sometimes used. One microtesla equals 10 milligauss.

transformer. The part of the electric power system that changes the voltage on power lines, usually "stepping it down" from the transmission lines to the lower voltages used on distribution lines.

transmission line. A power line carrying high-voltage electricity between regions. Most transmission lines are built on steel towers and operate at voltages between 60 and 765 kV.

V/m. Volts per meter. Electric fields are measured in volts per meter or in thousands of volts per meter (kV/m).

voltage. A measure of the electric tension or potential on a power line. Voltage is measured in volts (V).

wire code. A surrogate measure of magnetic fields based on the thickness, configuration and distance of nearby power lines.

x-rays. A form of ionizing electromagnetic radiation known to break chemical bonds.

Index

Adey, Ross, 52, 53, 73, 157, 161, 196

Alcedo, Juan, 142

Alternating current, 21, 37

Altoonian lawsuit, 111

Amador, 20, 21, 22, 23, 24

Amateur Radio Today, 149, 167

American Industrial Hygiene Assocation Journal, 143

American National Standards Institute, 148, 157

ANSI/IEEE safety limits, 148, 149, 157 - 159

Appliances, 120 - 125, 205

Association of Home Appliance Manufacturing, 121

Architecture, 140

Barry, John, 101, 156

Baseboard heating, 123

Bayliss, David, 63, 65, 66, 169

Beauregard, Donald, 17

Becker, Robert, 50, 51, 53, 175

Berryman, Patricia, 17, 18, 19

Bierman, David, 106, 107

"Biological Effects of Power Line Fields," 55

Body Electric, The, 51

Bone-growth stimulators, 51

Boston Globe, The, 64, 137

Bowman, Joseph D., 69, 105, 169

Breast cancer, 77, 86

Bromley, F. Allan, 63

Bryant, Arthur, 115

Bullock, Melissa, 116

Calcium efflux, 85, 86

California Department of Health Services EMF Program, 102

California State Department of Education EMF guidelines, 102

Canadian Center for Occupational Health and Safety, 136

Cancer, 81, 82, 86

Carcinogenesis, 82, 83

Carpenter, David O., 17, 57, 58, 66, 78, 79

Cell phone lawsuits, 159 - 160

Index

Cellular Technology Industries Association (CTIA), 162, 167

Cell phones, 162, 163

Chatkoff, Marvin L., 101

Circadian rhythms, 86

Circuit, 35

Citizens Concerned About EMFs, 19

Cleary, Stephen F., 81, 82, 161

Cleveland, Robert, 161

Clock, 125, 126

Cofactor, 91

Communication Workers of America (CWA/CIO), 147, 148, 149

Concrete-slab coil heaters, 126

Conductors, 44

Confounder, 90, 91

Contact current, 39

Corona, 37

Covalt lawsuit, 115

Crandall, Geoffrey, 174, 175

Criscuola lawsuit, 115

Current, 40, 127

Demand-side management (DSM), 174

Detroit Edison, 146

Dialectric union, 134

Direct current (DC), 37, 42

Distribution lines, 33, 130

Dosimeters, 105

Edison, Thomas, xv, 32

Electrical workers, 144 - 152

Electric blankets, 124, 125, 126

Electric charges, effects of, 36

Electric fields (EFs), 38

Electric Power Research Institute (EPRI), 66, 129, 130, 131

Electric power system, 32 - 36

Electric stoves, 127, 128

Electrifying America, 33

Electromagnetic fields (EMFs), 38

Electromagnetic Radiation Advisory Council, 156

Electromagnetic radiation, 38

Electromagnetic spectrum, 41 - 43

ELF (extremely low-frequency radiation), 40

EMF litigation, 60, 61, 114 - 117

Empire State Electric Energy Research Corporation (ESEERCO), 140

Enertech, 22

Epidemiology, 91

Ewing, Frank, 118

WARNING: the electricity around you may be hazardous to your health

Federal Communications Commission, (FCC), 158, 159
Federal Telecommunications Act, 30, 164
Filipowski, John, 26, 27
Florida Today, 166
Florig, H. Keith, 58
FM radio, 43
Frequency, 43, 44
Fresno Unified School District, 17

Gauss (G), 40
Gaussmeter, 40, 103, 104, 105, 207
Generators, 32, 33
Ghandi, Om, 160
Gillette, Lynne, 103, 124
Gillespie, Mary, 123
Glazier, Eve, 17
Goldstein, Norman N., 175
Goodman, Reba, 81
Greenberg, Mel, 24, 25
Gurda, Michael, III, 27, 28, 29

Hair dryers, 123
Halper, Martin, 64
Ham radio operators, 149
Henderson, Ann, 81
Hertz, 38

High-current configuration house, 129
High-Voltage Transmission Research Center, 110
Hill, Doreen, 60, 102
Hocking, Bruce, 162
Home appliances (see appliances)
Home grounding system, 130, 131
Hormone production, 82
Houston Lighting and Power, 57, 58
Houston litigation, 57, 58
Houton, Mike, 159, 162, 163
Hutchison, George, 63
Hydroelectric power stations, 33, 36

Iannucci, Barbara, 24, 25
Institute of Electrical and Electronic Engineers (IEEE), 149, 150
    RF Standard, 157 - 159
International Brotherhood of Electrical Workers (IBEW), 144
Ionizing radiation, 43, 79

Jersey Central Power and Light (JCPL), 24, 25, 26

Kaiser Permanente Hospital, Oakland, 135
Kane v. Motorola, 159
Kaufman, Herbert M., 140, 176

Index

Klein Independent School District, 57, 58
Knob and tube wiring, 132
Kreutzer Richard, 22, 23
Kues, Henry, 152, 161

Lai, Henry, 161, 162
Lee, William, 155
Leeper, Ed, 50
LeGrand, Dave, 144
Linet Study, 69, 70
Luben, Richard, 68, 69

Magnetic field, 39, 40
Magnetic permeability, 45
Magnetometer, 40
Mantiply, Ed, 155, 166
Marcy-South line, 26 - 29
Marino, Andrew, 25, 26, 48, 56
Matanowski, Genevieve, 58, 59
Maurer, Stewart, 115
McGill University Workers Study, 143
Measuring magnetic fields, 117 - 129, 138
Medina, City of, 29, 30, 31
Melatonin, 83
Microwave Debate, The, 155

Microwave ovens, 125, 159
Microwave radiation sickness, 152, 153, 163, 167
Middletown, New Jersey, 23, 24
Milham, Samuel, 167
Mitigation, methods, 109, 110, 111
Modulated RF radiation, 153
Montecito, California, 22, 23
Montecito Union School, 23
Monte Video School, 20
Moore, Catherine, 103
Morgan, M. Granger, 59, 86
Motorola, 162, 163
Murley, Clyde, 92
Mutagenic effects, 78 - 81

Nair, Indira, 59, 96, 97, 98
NAS/NRC Review, 68, 69
National Cancer Institute (NCI) Study, 69, 70
National Council on Radiation Protection (NCRP), 158
National Electric Safety Code (NESC), 131
National Electromagnetic Field Testing Association, 103
National Institute for Occupational Safety and Health (NIOSH), 145, 146, 151
Natural killer (NK) cells, 79

WARNING: the electricity around you may be hazardous to your health

Neurotransmitters, 83
Neutra, Raymond P., 22, 156
New England Journal of Medicine, 69
New Jersey Commission of Radiation Protection, 100, 122
New Jersey Public Advocate, 24
New York Power Authority (NYPA), 52
New York State Department of Health, 52
New York State Labor Institute, 145
New York State Power Lines Project Study, 53-57
New York Telephone Company, 59, 144
NIOSH safety recommendations, 146
Nixon, Hal, 143, 144
Noe valley cancer cluster, 156
Nonionizing radiation, 42, 43
Norton, Joe, 107
Nye, David E., 33

Occupational Safety and Health Administration (OSHA), 149
ODC, 83
Office of Technical Assessment (OTA), 58
Oge, Margo, 157
O'Malley, Terrance, 134

Pacemaker interference by cell-phones, 163
Pacific Gas and Electric (PG&E), 17-20, 171
Pacific Northwest Laboratory, 83
Patrick AF Base radars, 166, 167
Peak usage, 119
Peters, John, 65
Peters, Rosemary, 27, 28
Phillips, Jerry, 45, 53, 57
Photocopiers, 139
Police radar guns, 150
Poole, Charlie, 64
Portable electric coil heaters, 123
Project Sanguine (1973), 46, 47, 50
Prudent avoidance, xiii, 96, 111
Public Utilities Commission (Public Services Commission), 107

Radar, 165
Radio-frequency radiation (RF), 43, 152
Rauch, Greg, 129
Reiter, Russell, 83
Repacholi study, 162
Reproductive hazards, 86
Residents Against Giant Electric (RAGE), 24, 25
Reynard v. NEC, 159

Index

RF/microwave workers, 147 - 149
RF standard, 148, 157
Riley, Karl, 18, 99, 103
RNA messenger molecules, 79
ROWs, 107, 108

Sage, Cindy, 103, 104, 168
San Diego Power and Light, 129
San Francisco Public Utilities Commission, 155, 165
San Francisco, RF scenarios, 155
San Ramon, 22, 23
Savitz, David, 52, 53, 59
Scientific method, 85 - 87
Selvin, Steve, 156
Service drop, 119
Shavers, electric, 123
Silva, Michael, 22
60-Hz magnetic field measurement data sheet, 119
Slater School, 15 - 20
Spring spectrum v. Medina, 30, 31
Stealth antennas, 165
Steneck, Nicholas H., 155
Stevens, Richard G., 74, 83
Stereos, 123
Stertzer, Fred, 100, 123
Strom, Robert S., 61, 152, 153

Substations, 32
"Suggested Protocol for Measuring 60-Hz Magnetic Fields in Residences," 118
Surrogate measures, 89
Sutro Tower, 155, 156
Sussman, Stan, 94
Swedish National Energy Administration, 100, 101
Swicord, Mays, 104

Television sets, 126
Tell, Richard, 134, 139
Tenforte, Tom, 159
Tesla, Nikola, 32
Tesla (t), 40
Trains, EMFs, 103
Transformers, 33
  pole mounted, 35
Transmission components, 33 - 35
Transmission Line Handbook, 105
Trial Lawyers for Public Justice (TLPJ), 113
Tumor growth, 78 - 80
Two-way switch, 132

Underground lines, 104
University of Texas Health Sciences Center, 83

Utility Workers of America, 143

US Environmental Protection Agency (EPA), 51, 59 - 61, 77, 157

Video arcades, 99

Video display terminals (VDTs), 134 - 139

Volt, 37

Voltage, 37

Wall wiring errors, 131

Ward, John, 112, 113

Wartenberg, Daniel, 100

Washington State EMF Task Force, 109

Waterbed heaters, 123

Water pipes and culverts, 130, 131

Wavelength, 42, 43

Wertheimer, Nancy, 49, 50

Wilson, Barry, 83

Wines, Kirk, 28, 29

Winters, Wendell, 53, 75

Wireless communication devices, 160

Wireless Technology Research Group (WTRG/SAG), 163

Wiring, unusual, 131

Wiring code configuration protocol, 129

Withey, Michael, 111, 153

Wolpe, Howard, 173

Wright v. Motorola, 160

Yannon v. RCA, 148, 152

Zafanella, Luciano, 93, 110

Zaret, Milton, 166, 167

To order
WARNING: The Electricity Around You May Be Hazardous to Your Health

send a check or money order for $16.95, plus $3.45 shipping and handling (per order) to:

Miriam PRESS
P.O. BOX 19-0936
MIAMI BEACH, FLORIDA 33119*

To pay with a major credit card, please fill in the information below:

Name: _____

Address: _____

City: _____ State: _____ Zip: _____

Charge to: ❑ American Express ❑ VISA ❑ MasterCard

Account No.: _____ Expir. Date: _____

Signature: _____

For information, please call us at (800) 884-6763

*Florida residents, please add $1.10 per book for Florida sales tax.